Metal Electrodes for Battery Technologies

Metal Electrodes for Battery Technologies

Edited by

Yurong Ren

School of Materials Science and Engineering, Changzhou University, Changzhou, Jiangsu 213164, People's Republic of China

Jianmin Ma

School of Materials and Energy, University of Electronic Science and Technology of China, Chengdu 611731, People's Republic of China

IOP Publishing, Bristol, UK

ISBN 978-0-7503-4871-3 (ebook)
ISBN 978-0-7503-4869-0 (print)
ISBN 978-0-7503-4872-0 (myPrint)
ISBN 978-0-7503-4870-6 (mobi)

DOI 10.1088/978-0-7503-4871-3

Version: 20211101

IOP ebooks

British Library Cataloguing-in-Publication Data: A catalogue record for this book is available from the British Library.

Published by IOP Publishing, wholly owned by The Institute of Physics, London

IOP Publishing, Temple Circus, Temple Way, Bristol, BS1 6HG, UK

US Office: IOP Publishing, Inc., 190 North Independence Mall West, Suite 601, Philadelphia, PA 19106, USA

Contents

Preface

Rechargeable batteries take us into a wonderful world. They can provide us with a convenient living means enriched by portable electronics, and environmentally friendly transportation with electric vehicles. As a result, rechargeable batteries have become a necessary part in our life. With the increase of their applications, rechargeable batteries with advanced performance are urgently needed. As one of important strategies, it is useful to employ metal materials (Li, Na, K, Mg, Al, Zn, Ge, Sn, Sb and Bi) as the anode for rechargeable (Li, Na, K, Mg, Al and Zn) batteries, since metal anodes exhibit low voltage and high capacities. Thus, rechargeable batteries with metal anodes has become a hot issue.

As one important part, Li, Na, K, Mg, Al and Zn metals exhibit high capacities in their corresponding metal batteries. Moreover, alloyed element anodes, i.e., Ge, Sn, Sb and Bi, are also a type of high-capacity anode for Li, Na and K-ion batteries. Therefore, we believe that it is necessary to write this book, which includes Li, Na, K, Mg, Al, Zn, Ge, Sn, Sb and Bi anodes. We introduce the properties of metal anodes, the development of metal anodes and challenges in batteries. This will demonstrate the overall development of metal electrodes for rechargeable batteries.

In order to finish this book, we invited some experts to write the chapters. Prof. Xiaochuan Duan from Taiyuan University of Technology, Prof. Yurong Ren from Changzhou University and Prof. Libao Chen from Central South University write the first chapter, which deals with the Li anode for Li metal batteries. Prof. Yong Liu from Henan University of Science and Technology and Prof. Jianmin Ma write the second chapter, which includes Na, K, Mg, Al and Zn anodes for their corresponding metal batteries. The third chapter on the Ge anode for Li, Na, K-ion batteries is written by Prof. Jun Wang from Shandong University. The fourth chapter on the Sn anode for Li, Na, K-ion batteries is written by Prof. Zhijia Zhang from Tiangong University. Sb and Bi anodes are finished by Dr Lei Zhang from Griffith University and Prof. Dan Li from Zhengzhou University, respectively. We hope that this book will help students from all levels and researchers to learn the metal electrodes for battery technologies.

Acknowledgement

We would like to express our sincere gratitude to Dr Gaoxue Jiang, Dr Yulu Yang, Dr Xi Tang, Mr Zhongsheng Wang, Mr Kanglong Guo for doing some editing work.

Editors biographies

Yurong Ren

Yurong Ren is now a professor in the School of Materials Science and Engineering at Changzhou University, Changzhou, China. She received her bachelor's degree from Jilin Normal University in July 1998 and PhD degree in Chengdu Institute of Organic Chemistry, Chinese Academy of Sciences in July 2010. She once worked in the University of California, Los Angeles as a visiting scholar from 2015 to 2016. She severs as the editor of Chinese Chemical Letters. Her main research topics include new carbon materials, new energy storage materials and related device systems. She has published more than 35 high-level research papers and she is a principal inventor of 11 Chinese patents.

Jianmin Ma

Jianmin Ma is a professor in University of Electronic Science and Technology of China, Chengdu, China. He received his BS degree in Chemistry from the Shanxi Normal University in 2003 and PhD degree in Materials Physics and Chemistry from the Nankai University in 2011. During 2011–2015, he also conducted research in several overseas universities as a postdoctoral research associate. He serves as the Academic Editor for Rare Metals, the Associate Editor for Chinese Chemical Letters, and editorial board member for Journal of Energy Chemistry, Nano-Micro Letters, Journal of Physics: Condensed Matter, and others. His research interest focuses on the energy storage devices and components including metal anodes and electrolytes, and theoretical calculations from density functional theory and molecular dynamics to finite element analysis.

List of contributors

Libao Chen
State Key Laboratory of Powder Metallurgy, Central South University, Changsha, Hunan 410083, People's Republic of China

Xiaochuan Duan
College of Materials Science and Engineering, Taiyuan University of Technology, Taiyuan 030024, People's Republic of China

Kaijia Feng
School of Materials Science and Engineering, Henan University of Science and Technology, Luoyang 471023, People's Republic of China

Yong Jiang
State Key Laboratory of Separation Membrane and Membrane Processes, Tianjin Municipal Key Laboratory of Advanced Fibers and Energy Storage, School of Material Science and Engineering, School of Mechanical Engineering, Tiangong University, Tianjin 300387, People's Republic of China

Dan Li
Centre for Catalysis and Clean Energy, Gold Coast Campus, Griffith University, Queensland 4222, Australia

Yong Liu
School of Materials Science and Engineering, Henan University of Science and Technology, Luoyang 471023, People's Republic of China

Jianmin Ma
School of Materials and Energy, University of Electronic Science and Technology of China, Chengdu 611731, People's Republic of China

Pingfan Ning
School of Electronic and Information Engineering, Tiangong University, Tianjin 300387, People's Republic of China

Zhiheng Pang
School of Mechanical Engineering, Tiangong University, Tianjin 300387, People's Republic of China

Han Qian
School of Materials Science and Engineering, Henan University of Science and Technology, Luoyang 471023, People's Republic of China

Yurong Ren
School of Materials Science and Engineering, Jiangsu Collaborative Innovation Center of Photovoltaic Science and Engineering, Changzhou University, Changzhou 213164, People's Republic of China

Heyi Sun
State Key Laboratory of Separation Membrane and Membrane Processes, Tianjin Municipal Key Laboratory of Advanced Fibers and Energy Storage, School of Material Science and Engineering, Tiangong University, Tianjin 300387, People's Republic of China

Huaihu Sun
School of Physics and Electronics Hunan University, Changsha 410082, People's Republic of China

Kai Sun
State Key Laboratory of Separation Membrane and Membrane Processes, Tianjin Municipal Key Laboratory of Advanced Fibers and Energy Storage, School of Material Science and Engineering, Tiangong University, Tianjin 300387, People's Republic of China

Shihao Sun
State Key Laboratory of Separation Membrane and Membrane Processes, Tianjin Municipal Key Laboratory of Advanced Fibers and Energy Storage, School of Material Science and Engineering, Tiangong University, Tianjin 300387, People's Republic of China

Feng Tao
School of Materials Science and Engineering, Henan University of Science and Technology, Luoyang 471023, People's Republic of China

Jun Wang
Key Laboratory for Liquid-Solid Structural Evolution and Processing of Materials, Shandong University, Jinan, Shandong 250061, People's Republic of China

Wenqiang Wang
School of Mechanical Engineering, Tiangong University, Tianjin 300387, People's Republic of China

Gang Xie
State Key Laboratory of Separation Membrane and Membrane Processes, Tianjin Municipal Key Laboratory of Advanced Fibers and Energy Storage, School of Material Science and Engineering, Tiangong University, Tianjin 300387, People's Republic of China

Shenglong Yan
School of Mechanical Engineering, Tiangong University, Tianjin 300387, People's Republic of China

Zhenyang Yu
School of Mechanical Engineering, Tiangong University, Tianjin 300387, People's Republic of China

Xiaoliang Zhai
School of Materials Science and Engineering, Henan University of Science and Technology, Luoyang 471023, People's Republic of China

Lei Zhang
Centre for Catalysis and Clean Energy, Gold Coast Campus, Griffith University, Queensland 4222, Australia

Mengmeng Zhang
State Key Laboratory of Separation Membrane and Membrane Processes, Tianjin Municipal Key Laboratory of Advanced Fibers and Energy Storage, School of Material Science and Engineering, School of Mechanical Engineering, Tiangong University, Tianjin 300387, People's Republic of China

Yang Zhang
State Key Laboratory of Separation Membrane and Membrane Processes, Tianjin Municipal Key Laboratory of Advanced Fibers and Energy Storage, School of Material Science and Engineering, Tiangong University, Tianjin 300387, People's Republic of China

Zhijia Zhang
State Key Laboratory of Separation Membrane and Membrane Processes, Tianjin Municipal Key Laboratory of Advanced Fibers and Energy Storage, School of Material Science and Engineering, School of Mechanical Engineering, Tiangong University, Tianjin 300387, People's Republic of China

IOP Publishing

Metal Electrodes for Battery Technologies

Yurong Ren and Jianmin Ma

Chapter 1

Lithium metal anodes for high-performance rechargeable batteries

Xiaochuan Duan, Yurong Ren and Libao Chen

Because of its extremely high theoretical specific capacity (3860 mAh·g^{-1}) and low redox potential (relative to standard hydrogen electrode of −3.040 V), metallic lithium is considered to be one of the promising choices for the next generation of anode materials for high-energy-density lithium metal batteries (LMBs), which can meet the rigid demands of new industries. However, lithium anodes have a series of bottlenecks such as low coulombic efficiency, poor cycle performance, and poor safety that need to be resolved. The electrochemical deposition of metallic lithium includes multiple steps such as ion transport and lithium nucleation growth. Understanding the law of lithium deposition and adjusting the metal deposition behavior on this basis are vital to the practical use of lithium metal anodes. This chapter summarizes the research progress of the lithium metal anode from the perspectives of lithium nucleation, lithium dendrite growth and interface stability, introduces the driving force of lithium nucleation, the heterogeneous nucleation model, the space charge model, etc, and the related modification strategies such as electrolyte modification, artificial solid electrolyte interphase (SEI) layer construction, and host framework composite. Finally, the main challenges facing the commercial application of lithium metal anodes are pointed out, and its future development is prospected.

1.1 Introduction

1.1.1 Li anode properties

With the continuous consumption of fossil fuels and the rapid growth of energy demand, the development of efficient technologies for energy conversion and storage has been regarded as one of the most important challenges in the world today. As a new green energy storage technology, lithium-ion batteries (LIBs) have the advantages of high-energy density and long cycle life. However, after nearly 30 years of development, the energy density of LIBs based on electrochemical intercalation

doi:10.1088/978-0-7503-4871-3ch1

reactions (graphite as a negative electrode material) is approaching the theoretical limit, but it still cannot meet the rapid growth of contemporary society's demand for energy storage [1]. Limited by the theoretical specific capacity density (approximately 372 mAh g^{-1}) and volume capacity density (approximately 735 mAh cm^{-3}) of the graphite anode, the current commercial LIB is close to its theoretical limit (mass energy density and volume energy density are ~250 Wh kg^{-1} and ~680 Wh L^{-1}, respectively), which greatly restricts the further improvement of commercial LIB. Therefore, it is imperative to develop next-generation electrode materials with high safety and high specific capacity [2].

The lithium (Li) element is in the second period of main group I of the periodic table. It has the lowest density (0.54 g cm^{-3}) among all metal elements and the lowest negative electrochemical potential (−3.04 V vs. the standard hydrogen electrode). Therefore, Li metal is a promising anode material candidate for the design of rechargeable batteries with high theoretical specific capacities (3860 mAh g^{-1} and 2061 mAh cm^{-3}) [3–7]. Broadly speaking, energy storage systems with metallic lithium as the negative electrode can be collectively referred to as lithium batteries, which can be divided into lithium primary batteries and lithium secondary batteries. The research on lithium primary batteries began in the 1950s, which used lithium metal as the anode material and manganese oxide (MnO_2), $(CF)_n$, disulfoxide ($SOCl_2$) and other materials as the cathode materials. In 1976, Whittingham from Exxon Company invented the Li/TiS_2 battery based on the electrochemical reaction equation: $xLi + TiS_2 \leftrightarrow Li_xTiS_2$, and then realized the commercialization of lithium metal batteries. However, the serious safety problems caused by lithium dendrites hindered the development of LMBs [8–10]. As a result, the LIBs with 'rocking chair' working principle was invented, which replaced the lithium metal negative electrode with a graphite negative electrode, and achieved great success. The revival of lithium metal anodes originated from the pursuit of higher energy density because of its high theoretic special capacity [11]. Moreover, lithium metal anodes can be adapted to high specific capacity cathode materials and are ideal anode materials for the construction of next-generation metal lithium-based battery systems such as lithium–air (Li–air) batteries, lithium–sulfur (Li–S) batteries, and solid-state lithium batteries [12]. Figure 1.1 exhibits the energy density of an energy storage system based on what a lithium metal anode can achieve [13]. The gravimetric energy density and volumetric energy density of the LMBs are much higher than those of LIBs based on a graphite anode. However, the active lithium metal electrode often causes side reactions, leading to serious safety problems. At present, the main problems that limit the commercial application of lithium metal anodes are: (i) the uneven nucleation and growth of lithium metal during repeated plating and stripping processes, resulting in uncontrollable formation of lithium dendrites and dead lithium (inactive lithium) and even electrode powdering, which can lead to piercing the separator and causing internal short circuits and thermal runaway, thus causing severe safety issues [14]; (ii) unstable and thick solid electrolyte interphase (SEI) layers will be formed on the Li metal surface due to the thermodynamic instability of Li metal, which can cause irreversible and continuous consumption of Li and electrolyte and increasing the internal resistance, thereby reducing the

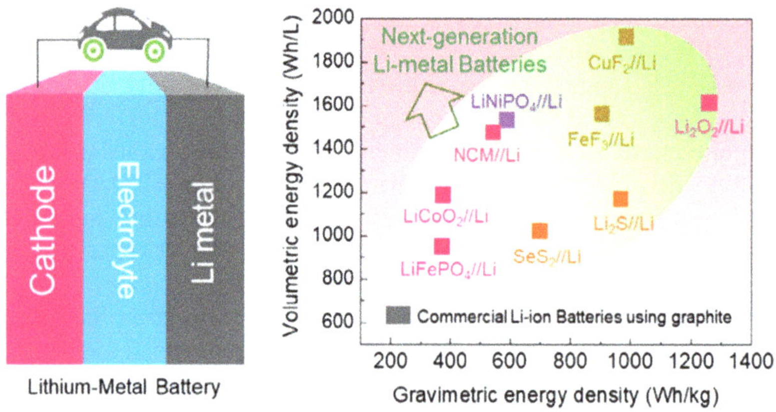

Figure 1.1. Schematic illustration of cell structure for LMB and the practically achievable energy densities of LMBs using different cathode materials. Source: [13]. Reproduced with permission of Elsevier B. V. and Science Press.

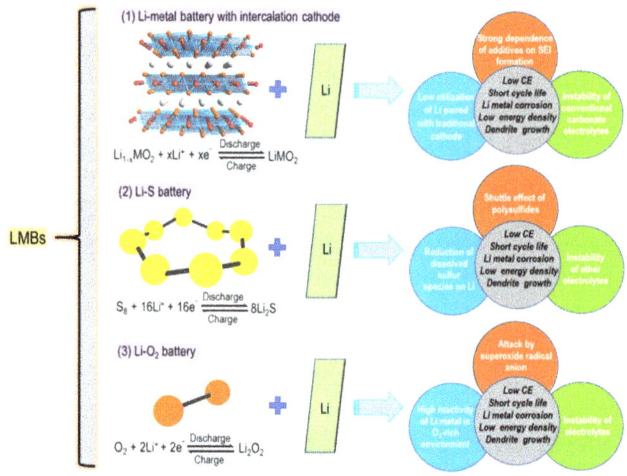

Figure 1.2. Three types of typical LMBs and their challenges facing lithium metal anode. Source: [16]. Reproduced with permission of Elsevier Inc.

coulombic efficiency and shortening cycle life [15]. Figure 1.2 summarizes the three most widely studied types of typical LMBs and their challenges facing lithium metal anode [16]. In particular, for Li–S and Li–air batteries, Li metal anode often also experiences further attack from polysulfide anions in Li–S batteries and from superoxide radical anions in Li–air batteries, which should be paid attention to in the design of an ideal lithium metal anode. It should be noted that the stable operation of Li metal anodes is prerequisite for achieving high-energy density batteries. Therefore, an in-depth understanding of the working principle and failure mechanism of lithium metal anodes are crucial for designing effective protection strategies and preparing high-performance LMBs.

1.1.2 Failure mechanism of Li metal anode

In order to improve the cycle performance and overcome the above safety issues for LMBs, it is critical to further understand the failure process of the lithium metal anode [17]. Due to the high activity of the lithium element, a few impurities and structural defects are inevitably formed on the surface of the metal lithium electrode during the preparation and storage processes of lithium metal electrodes, resulting in uneven local current density at the interface in the lithium deposition process. In the early stage of lithium deposition, lithium tends to preferentially nucleate at active sites with relatively high current density, which triggers the formation of moss-like deposition morphology, and subsequently leads to the growth of lithium dendrites [18–20]. In the following lithium dissolution process, lithium is more likely to dissolve preferentially at the newly formed dendrite sites due to the high activity and low impedance, thereby reducing the bonding force between the dendrites and the electrode [21]. Under the stress caused by the volume change, some of the weak dendrites will fall off and be wrapped by the SEI film to form electronically inert 'dead lithium' particles [22–24]. As the charge–discharge cycle progresses, the dead lithium particles on the surface of the lithium electrode accumulate with each other to form a dense insulating layer, which hinders the diffusion of lithium ions and the transmission of electrons. This will undoubtedly cause the polarization of the internal electrode reaction to increase, and eventually leading to the failure of LMBs. Therefore, an in-depth understanding of the formation and growth mechanism of lithium dendrites, the characteristics and changes of the SEI film on the surface of lithium metal, and the formation mechanism of 'dead lithium' are key to recognizing the failure of lithium metal anodes.

1.1.2.1 Formation mechanism of lithium dendrites

The problems of low safety and poor cycle performance caused by the growth of lithium dendrites are the main reasons that hinder the development of metal lithium anodes in scientific research and industrialization. At present, there are many research reports on the formation mechanism of lithium dendrites and their growth behavior hypotheses, including the nucleation/diffusion model, heterogeneous nucleation model, and space charge model, which provides a more detailed and solid theoretical basis to solve the problem of lithium dendrites [25].

1.1.2.1.1 Initial stage of nucleation

During the charging and discharging process, the lithium metal anode will undergo repeated electrochemical plating and stripping. During the deposition process, lithium ions are reduced to lithium atoms, and further gather to form small crystal nuclei, which is the nucleation process of lithium [26]. This process is driven by electromotive force and is usually determined by the combined forces of charge transfer overpotential, diffusion overpotential, and surface/interface formation overpotential. Generally, the nucleation process is accompanied by an increase in surface free energy and a decrease in volume free energy. The stability of the lithium nucleus depends on the change in total free energy, which can be described by the classic nucleation equation involving the relationship among the size, overpotential, and current density of lithium nuclei [27].

Assuming that the deposited nucleus is a spherical cap with a radius of r (figure 1.3) [28], the corresponding Gibbs free energy (ΔG_n) for this nucleation process is the sum of the volume free energy and the surface free energy:

$$\Delta G_n = -\frac{4}{3}\pi r^3 \Delta G_V + 4\pi r^3 \gamma_{NE} V_m$$

where ΔG_V is the volume free energy change, γ_{NE} is the surface energy of the nuclear/electrolyte interface, V_m is the molar volume of lithium. Accordingly, the critical radius (r_m) of lithium metal nucleation is:

$$r_m = \frac{2\gamma V_m}{F \mid \eta \mid}$$

where F is the Faraday constant, and η is the deposition overpotential. Therefore, the size of the lithium nucleus is inversely proportional to the nucleation overpotential, which is also related to the applied current of electrodeposition. The nucleation overpotential is large at high current density, while the overpotential is smaller at low current density.

Biswal *et al* systematically investigated the nucleation and early stage growth of Li electrodeposits [29]. According to the voltage–capacity profile of galvanostatic Li deposition, they proposed that the lithium deposition process can be divided into two stages (figure 1.4): (i) The initial potential drops sharply below 0 V (vs. Li^+/Li), and the spikes appearing in the potential curve indicate the nucleation process. At this time, lithium ions are driven to form lithium nuclei under the action of overpotential. This process needs to overcome the nucleation barrier, and the

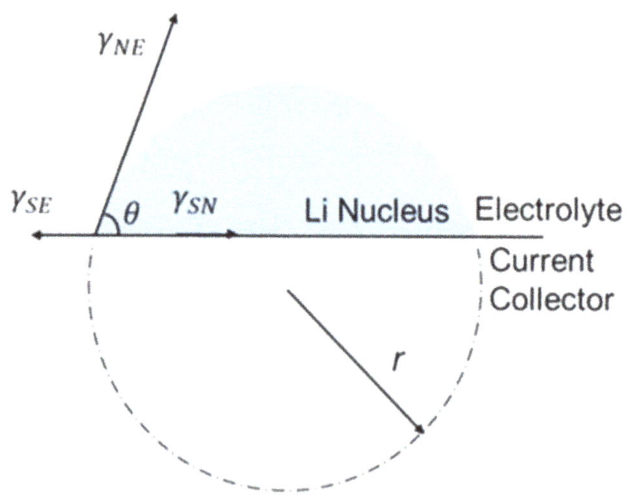

Figure 1.3. Schematic illustration of the formation of spherical lithium nucleus. Source: [28]. Reproduced with permission of China Association for Science and Technology.

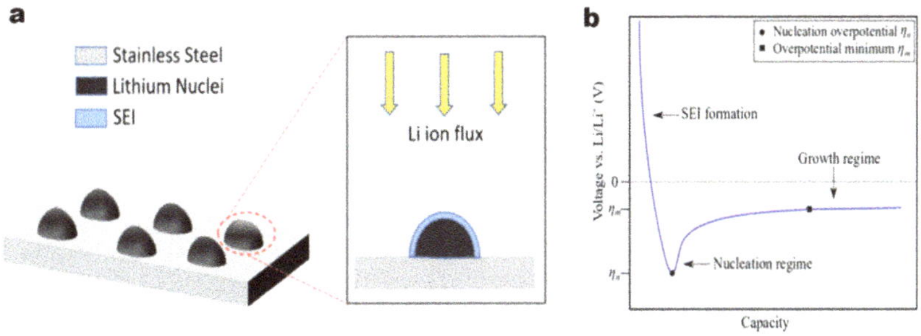

Figure 1.4. (a) Schematic illustration of the nucleation of Li ion by electrodeposition; (b) Voltage–capacity curve of galvanostatic Li deposition. Source: [29]. Reproduced with permission of American Chemical Society.

voltage at the peak is called the nucleation overpotential (η_n). (ii) After the initial formation of lithium nuclei, the potential begins to rise to η_m, which corresponds to the lithium growth stage after nucleation. The occurrence of η_m is mainly due to the need for Li^+ ion nucleation to reach the surface of the substrate through the SEI film. The absolute value of η_m is smaller than η_n, indicating that the energy barrier of lithium atom deposition on the formed lithium nucleus is lower, and lithium tends to grow on the existing nucleus instead of forming a new nucleus. The nucleation overpotential is the result of the overpotential during the charge transfer process and the overpotential during the formation of the lithium/solution interface [30–32]. In addition, the overpotential caused by concentration polarization is mainly affected by electrolyte diffusion. Generally speaking, at low current density, the formed lithium nuclei will be relatively large and sparsely distributed, while at high current density, the formed lithium nuclei are smaller and more densely distributed [33–35]. There are two common models used for describing the nucleation process: the heterogeneous nucleation model and the space charge model.

1.1.2.1.2 Heterogeneous nucleation model

When the substrate is non-metallic lithium, such as copper and iron, the electrodeposition process of lithium ions on the surface is considered to be a heterogeneous nucleation behavior, and the final deposition morphology is greatly affected by the initial nucleation process. Therefore, an in-depth understanding of the heterogeneous nucleation behavior of lithium is key to controlling the morphology of lithium deposition [36]. The critical radius theory is also applicable to heterogeneous nucleation model (figure 1.3), lithium nuclei continue to grow to a critical size and reach a stable state of thermodynamics and kinetics. When the nucleation of lithium deposition is relatively fast, the lithium nucleus generated by heterogeneous nucleation is morphologically unstable, and it is easy to form dendritic lithium [37–39]. In addition, lithium metal extrusion caused by stress will occur at the heterogeneous interface, which will also trigger the growth of lithium dendrites [25, 40–42]. According to the heterogeneous nucleation theory, the relationship between the critical free energy ΔG^*_{eq} and the contact angle can be described as:

$$\cos\theta = \frac{\gamma_{SE} - \gamma_{SN}}{\gamma_{NE}}$$

$$\Delta G_{eq}^{*} = \frac{1}{3}\left(4\pi r_{eq}^{*2}\right)\gamma_{NE}\left(\frac{2 - 3\cos\theta + \cos^{3}\theta}{4}\right)$$

where θ is the contact angle between the Li nucleus and the electrolyte, and γ_{SE}, γ_{SN}, and γ_{NE} are the free energy of the interface of the substrate/electrolyte, substrate/nucleus and nucleus/electrolyte, respectively. When the contact angle $\theta = 0°$, the free energy of nucleation is 0; when $\theta = 180°$, the required free energy of nucleation is the same as the energy required to achieve uniform nucleation. By improving the wettability of the lithium nuclei and the substrate, the contact angle can be reduced, thereby reducing the free energy of nucleation. Therefore, the properties of the substrate have an important influence on the nucleation behavior of lithium.

Cui and co-workers systematically compared the heterogeneous nucleation growth process of lithium on different substrates [38]. As shown in figure 1.5(a), when a copper foil with a smooth surface is used as a current collector, an obvious voltage drop appears on the capacity curve in the initial lithium deposition, and then the voltage rises and presents a stable plateau. Here, the tip value of the voltage drop can be regarded as the initial nucleation overpotential for the lithium deposition. Due to the large thermodynamic mismatch between the lithium and copper substrate, an overpotential of about 40 mV is needed to overcome the barrier of heterogeneous nucleation. In contrast, when the gold substrate is used, two lithiation voltage plateaus appear above 0 V, corresponding to the formation of Li_xAu alloy (figure 1.5(b)). Afterwards, below 0 V, the voltage–capacity curve of lithium deposition on the surface of the gold substrate shows a corner signal, indicating that the overpotential required for the initial nucleation of lithium on the Li_xAu alloy is almost zero, which is distinct from the copper substrate. Therefore, when the lithium deposition capacity is limited, lithium is more likely to be deposited on the surface of the gold substrate than the copper substrate. In summary, the heterogeneous nucleation process of lithium can be effectively controlled by modifying the lithium deposition substrate or current collector, thus the localized growth and uniform deposition of metallic lithium can be further realized.

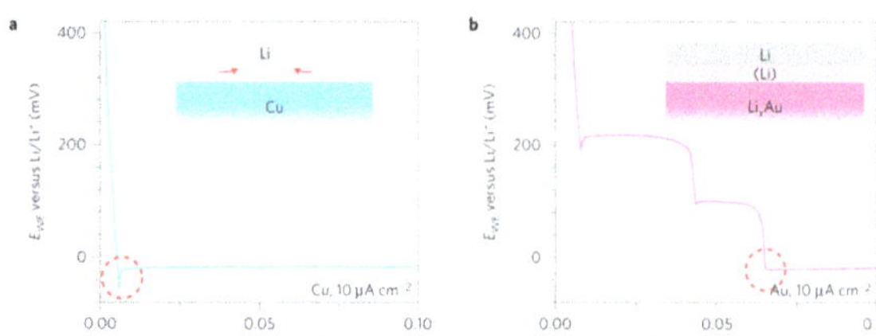

Figure 1.5. Overpotential during galvanostatic Li deposition on (a) Cu and (b) Au substrates. Source: [38]. Reproduced with permission of Nature Publishing Group.

1.1.2.1.3 Space charge model

After the initial nucleation is formed, the heterogeneous nucleation theory is no longer applicable to explain the growth behavior of lithium on the lithium surface. In order to explore the deposition mode after initial nucleation, Chazalviel proposed a space charge model, which is mainly suitable for low-concentration electrolytes or rapid lithium deposition processes [43]. Under the action of an electric field, the relatively dispersed lithium ions and anions move to the negative electrode and the positive electrode, respectively, resulting in a decrease of anions on the surface of the negative electrode. The depletion of anions on the electrode surface will generate space charges and electric fields near the electrode/electrolyte interface, affect the structure and morphology of the deposits, and induce dendrites during the lithium deposition process. Under high current density conditions, excessive positive charges will induce the formation of local space charges, making the negative electrode unstable and prone to problems such as uneven deposition and dendrites [44–46].

The space charge model can be also used to interpret the Li dendrite formation and growth behavior. The initial time (τ) of dendrite growth based on the space charge model is called Sand time, and the calculation formula is expressed as follows:

$$\tau = \pi D \frac{eC_0(\mu_a + \mu_{\mathrm{Li}^+})^2}{2J\mu_a^2}$$

where D is the bipolar diffusion coefficient, e is the charge, C_0 is the initial concentration of the electrolyte, J is the effective current density, and μ_a and $\mu_{\mathrm{Li+}}$ are the mobility of anions and Li^+, respectively. The short Sand time is easy to cause the growth of lithium dendrites due to the space charge effect. As the deposition capacity increases, the lithium dendritic layer on the electrode surface gradually diffuses to the electrode convection area, as shown in figure 1.6(a) [47]. The rapidly growing lithium dendrites will pierce the separator and cause short circuits inside the battery. When the current density is larger, τ is smaller, indicating that the battery needs less time for internal short circuit, as shown in figure 1.6(b). Therefore, reducing the effective current density or accelerating the Li^+ mobility can effectively

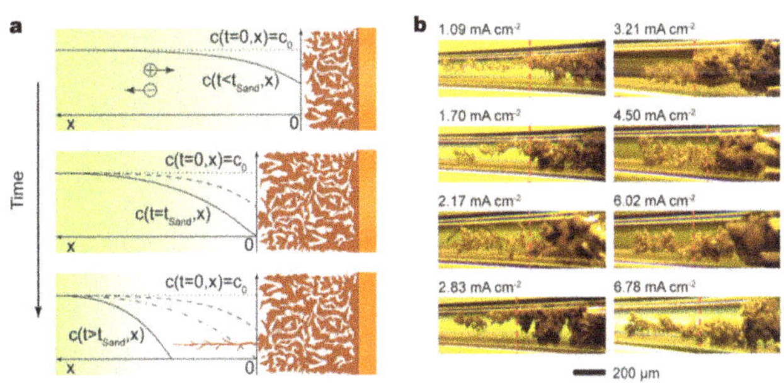

Figure 1.6. (a) Schematic illustration of the growth mechanism of electrodeposited lithium dendrites during concentration polarization; (b) The growth state of lithium dendrites under different deposition current densities. Source: [47]. Reproduced with permission of Royal Society of Chemistry.

extend the Sand time and delay the generation of lithium dendrites, thereby improving the cycle life and safety of the battery [48–50].

The heterogeneous nucleation model and the space charge model mainly explore the morphology of the lithium deposition process. Among them, the heterogeneous nucleation model is suitable for the initial stage of nucleation, and its nucleation behavior is affected by the substrate, such as crystal structure, lattice matching, crystal plane orientation and defects, etc; while the space charge model is suitable for the later growth stage of lithium, which is mainly affected by current density and electric field. However, the above two models also have certain limitations. For example, the factors including electrolyte decomposition and SEI formation during the deposition process are not considered, which will also affect the actual lithium nucleation and deposition behavior. The SEI film spontaneously formed on the surface of lithium metal has an important influence on the morphology of lithium deposition. During the charge–discharge cycle, the volume change effect of metallic lithium induces severe stress fluctuations at the electrode interface, leading to the rupture of the SEI film structure and exposing the highly active metallic lithium surface, which in turn induces uneven lithium deposition behavior and promotes the growth of spines lithium dendrites.

1.1.2.2 Formation of SEI layer

Due to the low reduction potential of organic solvents (< 1.0 V vs. Li/Li^+), the surface of the negative electrode exposed to the electrolyte will directly undergo an irreversible chemical reaction with the electrolyte components within a few milliseconds. The resulting electrode–electrolyte interface is called the solid electrolyte interface (SEI). The physical and chemical properties of the SEI film have a significant impact on the lithium metal anode [51–55]. Besides, the growth of lithium dendrites and the formation of dead lithium are also closely related to the morphology of the SEI film. Unlike other types of anode materials, the host-free lithium metal anode has significant interface structure evolution during charge and discharge. Therefore, when the structure of the SEI film on the surface of the lithium electrode is unstable, the SEI film is easily broken under the fluctuation of the internal stress of the electrode, thereby exposing a fresh and highly active metal lithium interface. These active interfaces will promote the reduction of the electrolyte to form a new SEI film, which will consume more electrolyte, and in severe cases will deplete the electrolyte. In addition, these newly exposed active regions lead to relatively high local current density, which will promote the growth of lithium dendrites, further aggravate the evolution of the interface structure, and form a vicious circle. At the same time, the cracked SEI film components continue to accumulate during the electrode cycle and accumulate on the electrode reaction interface, hindering the diffusion of lithium ions at the electrode interface, increasing the electrochemical polarization, and ultimately leading to the failure of the metal lithium electrode. Factors such as the distribution, composition, and structure of the SEI film will affect the deposition behavior of lithium, such as lithium nucleation overpotential, nucleation site distribution, and deposition morphology. Therefore, an in-depth understanding of the SEI film on the surface of lithium metal is key to improving the electrochemical performance of lithium metal anodes [56].

1.1.2.3 Formation of dead lithium layer

During the charging and discharging process, the lithium metal anodes have a significant volume change with the dissolution and deposition of lithium. The formation of the dead lithium layer on the surface of lithium electrode is related to the volume change of the electrode. As shown in figure 1.7, during the lithium deposition process, the electrode surface tends to form a mossy lithium deposition morphology due to the uneven nucleation and growth process [57]. During the initial delithiation, lithium is preferentially stripped out from the pre-existing mossy lithium. This is because compared to the bulk lithium, the moss-like lithium has higher activity and thus exhibits lower impedance characteristics. As the lithium extraction progresses, the size of the mossy lithium on the electrode surface is further reduced, causing the increase of internal resistance. Due to the volume change, these small-sized moss-like lithium will be mechanically separated from the electrode substrate under volumetric stress, thereby forming completely electronically inert dead lithium (the surface is wrapped by the SEI film). Subsequently, lithium is mainly extracted from the bulk lithium, causing a large number of small holes to be formed on the surface of the electrode, and these sites have strong activity and can serve as sites for lithium nucleation and growth, resulting in the formation of new mossy lithium. In addition, due to the tightly packed structure of the hard-shell battery, the dead lithium formed on the electrode surface is subject to greater mechanical compression stress, so that a dense inert dead lithium layer is easily formed at the electrode interface. The existence of the dead lithium layer largely hinders the diffusion of lithium ions on the electrode surface, which promotes the formation of lithium dendrites, and improves the polarization of the electrochemical reaction. After a long cycle, the growth of dead lithium leads to a decrease in active lithium, while the thickness and interface resistance of the SEI layer increase sharply, leading to battery failure.

1.2 Development of Li anode

Solving the failure problem of the metal lithium anode is the only way to realize the commercialization of LMBs. How to use the advantages of lithium electrodes and limit their disadvantages is a challenge that researchers in the field of lithium batteries must face. In view of the failure mechanism of the lithium metal anode,

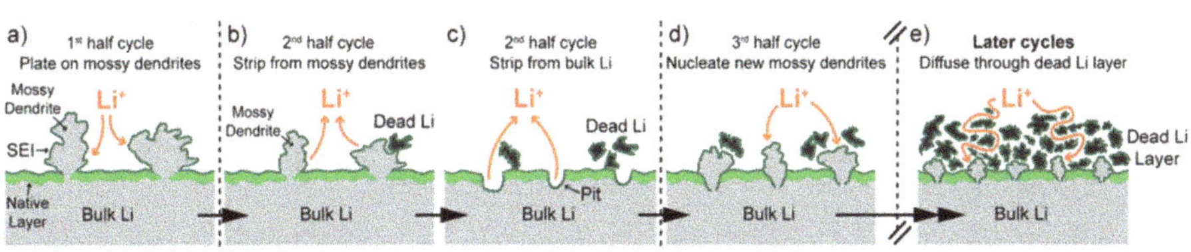

Figure 1.7. Schematic illustration of the forming mechanism of the dead lithium layer on the surface of lithium metal anode. Source: [57]. Reproduced with permission of Royal Society of Chemistry.

including the growth of lithium dendrites, the formation of SEI film and dead lithium layer, and the change of reaction interface structure, researchers have proposed a variety of strategies to solve the problem of lithium metal failure, including electrolyte modification, artificial SEI layer construction, host framework composite, etc, which have greatly improved the performance of the metal lithium anode.

1.2.1 Electrolyte additives

During the charging/discharging processes, the Li metal will react with electrolyte to form the SEI layer, which has an important impact on the Coulomb efficiency and dendrite growth of the lithium metal anode [58–63]. In traditional $LiPF_6$-based ester electrolytes, the Coulombic efficiency of lithium metal anode deposition and stripping is low, and the formation and growth of lithium dendrites is serious. The electrolyte chemistry can be adjusted by optimizing electrolyte additives, solvents, lithium salt, and lithium salt concentration. For example, the use of ether solvents that are more compatible with lithium metal anodes can reduce the reactivity of the electrolyte with lithium metal; the use of a variety of additives and new lithium salts can effectively inhibit the formation of lithium dendrites; the use of high-concentration lithium salt electrolyte can form a stable SEI film, etc. Among them, the electrolyte additive is a convenient strategy to improve the stability of the lithium metal anode since it does not need to redesign commercial battery components and structures, which has a unique advantage for the commercialization of lithium metal batteries. There are three main types of electrolyte additives that have been reported for LMBs based on their protection mechanism:

(i) Protective layer forming agent. The introduction of electrolyte additives can promote the formation of film on the electrode surface, and increases the stability of the electrode and electrolyte interface. The stability of the lithium metal anode can be improved by designing the additive functional group. Ma *et al* systematically studied the mechanism of potassium perfluorohexyl sulfonate (K^+PFHS) additive on the improvement of the electrochemical performance of NMC/Li full battery, as shown in figure 1.8 [64]. Due to its high fluorine content and the sulfonic acid functional group, K^+PFHS can coordinate with lithium ions. Theoretical study found that K^+PFHS has the lower lowest unoccupied molecular orbital (LUMO) and the higher highest occupied molecular orbital (HOMO). During the charging/discharging cycle, the K^+PFHS additive can be reduced before the electrolyte oxidize to form the SEI layer on the surface of the electrodes to inhibit the further decomposition of the electrolyte and improve the Coulombic efficiency of the battery. On the side of the lithium metal anode, the decomposition of K^+PFHS provides a large amount of fluorine, which increases the content of LiF in the SEI film. This is beneficial to reduce the thickness of the SEI film and increase the Li^+ ion migration rate. It is also found that a thin and uniform cathode electrolyte interphase (CEI) layer is formed on the $LiNi_{0.6}Mn_{0.2}Co_{0.2}O_2$ cathode (NMC) side, which not only

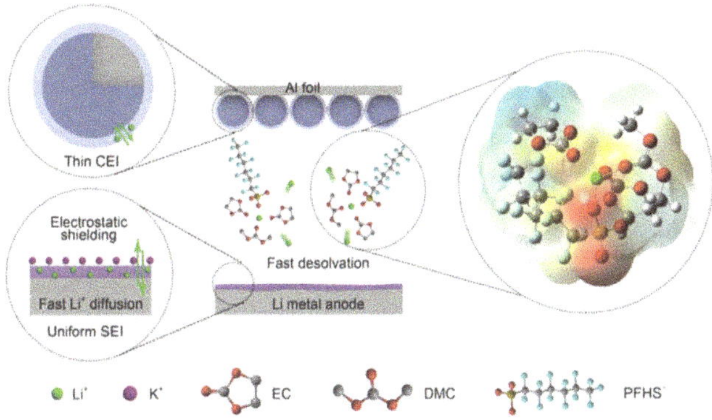

Figure 1.8. Schematic illustration of the stabilization mechanism of K^+PFHS additives at the electrolyte/cathode and anode interface. Source: [64]. Reproduced with permission of Elsevier B.V. and Science China Press.

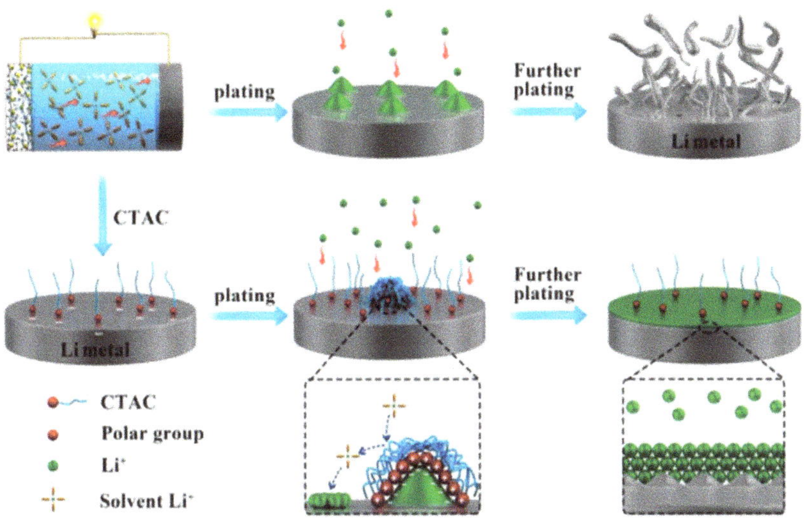

Figure 1.9. Schematic illustration of the lithiophobic repulsion effect of CTAC additive on the Li decomposition process. Source: [65]. Reproduced with permission of American Chemical Society.

shortens the Li^+ ion diffusion distance, but also stabilizes the structure of the cathode material. These changes improve the coulombic efficiency and cycle stability of LMBs.

(ii) Cation shielding agent. By introducing cations that can be preferentially adsorbed on the surface of the lithium nucleus, the growth of lithium dendrites can be effectively inhibited. Lai and co-workers reported a cation shielding electrolyte additive, hexadecyl trimethylammonium chloride (CTAC), which can effectively suppress the growth of lithium dendrites by lithiophobic repulsion mechanism, as shown in figure 1.9 [65]. Due to the electrostatic attraction, cationic surfactant molecules CTAC aggregate at the protrusions where lithium nuclei grow, forming a lithiophobic protective outer layer. This shielding effect can induce the deposition of Li^+ ions to adjacent regions, thus hindering the formation of lithium dendrites.

(iii) Solvation regulator. Lithium ions are always surrounded by solvent molecules to form a solvation shell in the electrolyte. However, in the electrochemical reaction process, lithium ions need to be removed from the solvation shell before the subsequent intercalation reaction or deposition reaction can proceed. Adjusting the solvation structure and reducing the energy barrier for Li^+ ion extraction cannot only accelerate the progress of the electrochemical reaction, but also prevent the solvation shell from being embedded in the electrode material and destroy the structure of the electrode material. In addition, the components of SEI are mainly derived from the solvated layer of lithium ions in the liquid electrolyte. Therefore, adjusting the Li^+ ion solvation layer can significantly improve the uniformity of SEI and inhibit the formation of lithium dendrites. Zhang *et al* introduced fluoroethylene carbonate (FEC) and lithium nitrate ($LiNO_3$) into the electrolyte at the same time through the solvation of the ether ester mixed solution, and found that the composition and structure of the Li^+ ion solvation layer can be effectively changed, resulting in a uniform SEI layer composed of LiF and LiN_xO_y with high content (figure 1.10), finally obtaining uniform lithium deposition morphology [66].

1.2.2 Artificial SEI layer construction

The artificial interface design is to construct an artificial SEI layer as interface protective layer on the surface of metal lithium through nanotechnology with high ionic conductivity, stable physical and chemical properties, good interface compatibility, and appropriate mechanical strength. The artificial SEI layer can effectively inhibit the side reaction of metallic lithium and electrolyte, and improve the plating/stripping efficiency of metallic lithium [67–73]. The interface compatibility adjustment can effectively reduce the interface impedance of the electrolyte, especially the solid electrolyte and metallic lithium, and improve the interface conductivity.

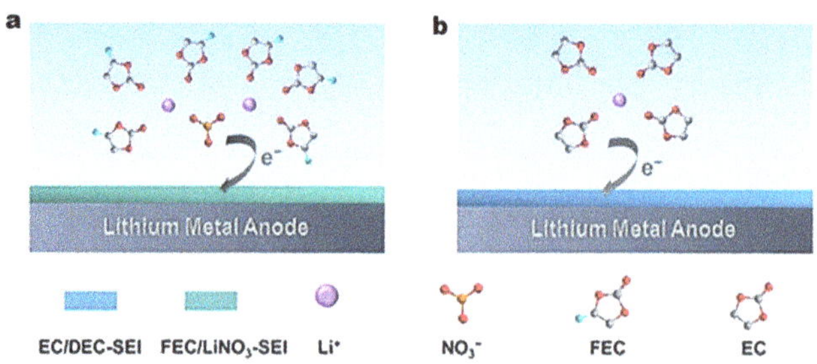

Figure 1.10. Schematic illustration of solvation sheath of lithium ions and the SEI formed in (a) FEC/$LiNO_3$ and (b) EC/DEC electrolytes. Source: [66]. Reproduced with permission of Wiley-VCH Verlag GmbH & Co. KGaA, Weinheim.

The design of artificial SEI layer can also guide the uniform nucleation growth of metal lithium and realize the uniform deposition of metal lithium, thus the growth of lithium dendrites can be inhibited since the flexible artificial SEI layer can adapt to the volume change of the metal lithium anode. Based on their composition, the artificial SEI layers can be divided into inorganic SEI layers, organic polymer SEI layers, and organic/inorganic composite SEI layers.

1.2.2.1 Inorganic SEI layers

The inorganic artificial SEI layers are generally composed by fast ion conductor ceramics (such as Li_3PO_4, Li_3PS_4, Li_3N, Li_2S, etc) that have a high Young's modulus and high Li^+ ion conductivity, which can inhibit the growth of lithium dendrites and significantly improve the ion transmission capacity of the interface. In addition, lithium halides, alumina, etc can stabilize the interface, greatly reduce side reactions, and improve interface compatibility [74]. The fast ion conductor inorganic artificial SEI layer can be obtained by *in situ* reaction between a suitable precursor and metallic lithium. Guo *et al* immersed lithium metal in a polyphosphoric acid (PPA) solution to form *in situ* a uniform and smooth Li_3PO_4-based SEI layer [73]. As shown in figure 1.11, the passivation layer on the surface of ordinary untreated lithium metal (the main components are $LiCO_3$, LiOH and Li_2O) will spontaneously react with the electrolyte to form an SEI layer, forming a porous structure on the surface of the lithium metal, resulting in the increase of local Li^+ ion flux, which in turn promotes the nucleation and growth of lithium dendrites. When the Li_3PO_4-based artificial SEI layer is introduced, the protective layer blocks the side reaction between lithium metal and the electrolyte. X-ray photoelectron spectroscopy (XPS) results show that Li_3PO_4 can exist stably after cycling. The interface protection layer has a high Young's modulus of 10–11 GPa, which can effectively inhibit the growth of dendrites [75].

1.2.2.2 Polymer SEI layers

Because of the convenient manufacturing process, excellent flexibility and good stability, the polymer artificial SEI layer can withstand unlimited volume changes during battery cycling, inhibit dendrite growth, and realize a lithium metal anode with high rate and long cycle life. Uneven Li^+ ion conduction and uneven

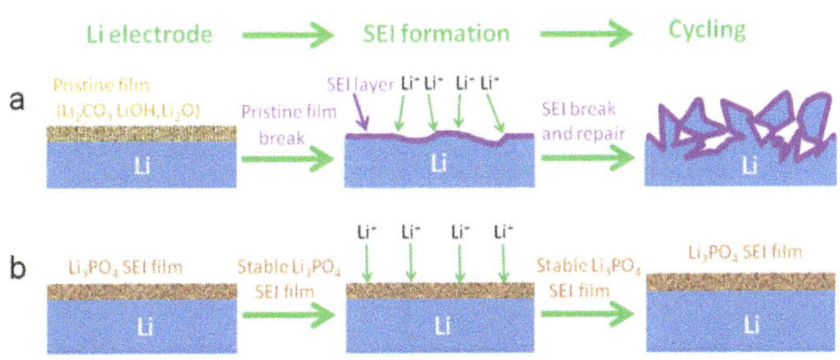

Figure 1.11. Schematics of (a) the general Li metal anode and (b) Li_3PO_4-modified Li metal anodes during SEI formation and cycling. Source: [73]. Reproduced with permission of Wiley-VCH Verlag GmbH & Co. KGaA, Weinheim.

distribution of charges on the electrode surface are the main reasons for the growth of lithium dendrites. Due to the strong interaction between Li^+ and polar groups (–COOR, –OH, –F, etc), using a polymer layer with polar groups as an artificial SEI film can effectively control the flux of lithium ions evenly distributed on the surface of lithium metal [76]. Based on the common polymer artificial SEI layer with different chemical properties and mechanical properties, Bao *et al* proposed the design and modification principles of the polymer artificial SEI layer (figure 1.12) [77]: (i) The thickness of the protective film has a significant impact on the lithium deposition process. An artificial protective film that is too thin will affect the lithium nucleation process, and too thick will limit the transmission of Li^+, thereby reducing the uniformity of lithium deposition. (ii) Self-healing polymers (SHP) and self-healing elastomers (SHE) have similar chemical properties, but SHP is a flowable viscoelastic material, and SHE is a covalently cross-linked elastic solid. By comparing SHP and SHE two polymer films, the mechanical properties of the polymer film will affect the uniform deposition of lithium ions, but its chemical functionality is the main factor affecting the local morphology of Li. (iii) As shown in figure 1.12(c), the dielectric constant and surface energy of the polymer together affect the size of the deposited particles. The exchange current increases with the increase of the dielectric constant of the polymer film, while the polymer with low surface energy will produce higher interfacial energy, thereby jointly increasing the particle size during Li^+ ion deposition.

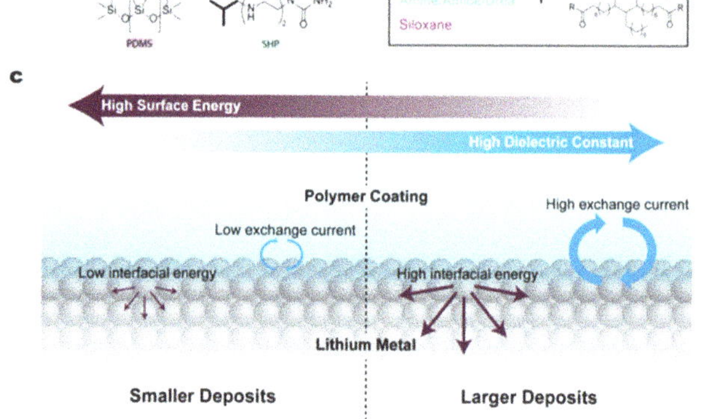

Figure 1.12. (a) The chemical structure of the polymer used to construct the artificial polymer SEI layer. (b) The schematic of the initial growth stage of lithium metal in the presence of the artificial polymer SEI layer. (c) The factors affecting the deposition of lithium metal by the artificial polymer SEI layer. Source: [77]. Reproduced with permission of American Chemical Society.

1.2.2.3 Organic/inorganic composite SEI layers

Inorganic and polymer artificial SEI films have certain limitations. For example, the polymer film can alleviate the volume change of the electrode during the cycle and promote Li^+ ion transport, but the effect of inhibiting lithium dendrites during long battery cycles is unsatisfactory due to its low mechanical strength. The inorganic artificial SEI film with high mechanical strength can make up for the shortcomings of the polymer film, but it is brittle and difficult to process. The organic/inorganic composite artificial SEI layer combines the flexibility of organic film and the advantages of high ionic conductivity and mechanical strength of inorganic matter. It cannot only promote the uniform diffusion of lithium ions on the electrode surface, but also effectively alleviate the electrode's cycling process. Through the synergistic effect of the organic and inorganic components, the growth of lithium dendrites is effectively inhibited, and the cycle life of the metal lithium battery is improved [78–80]. According to its component properties, the organic/inorganic composite artificial SEI layers can be divided into inorganic fast ion conductor/ organic composite artificial SEI layers, organic ionic conductor/inorganic composite artificial SEI layers, and multi-component composite artificial SEI layers.

The garnet-type inorganic electrolyte has good stability with metallic lithium, strong rigidity and inherent high ionic conductivity, thus it can be used as the inorganic component of the artificial composite interface layer combined with flexible organic polymers. Xu *et al* synthesized a two-phase single-ion conductive artificial composite interface layer, which is composed of rigid aluminum-doped garnet structure $Li_{6.75}La_3Zr_{1.75}Ta_{0.25}O_{12}$ (LLZTO) inorganic substance as the skeleton, and soft lithiated naphthene (Nafion) is constituted as a filler (denoted as LLN) [81]. The inorganic layer can maintain high mechanical stiffness and rapid Li^+ ion conductivity, and the lithiated naphthfine layer can give the interface layer greater elasticity to adapt to the volume change of the electrode. As shown in figure 1.13, LLN has single-ion conductivity (the Li^+ ion migration number reaches 0.82), which effectively blocks the passage of anions, so that Li^+ ions can diffuse uniformly and efficiently on the electrode surface, and promote the dense deposition of metal lithium.

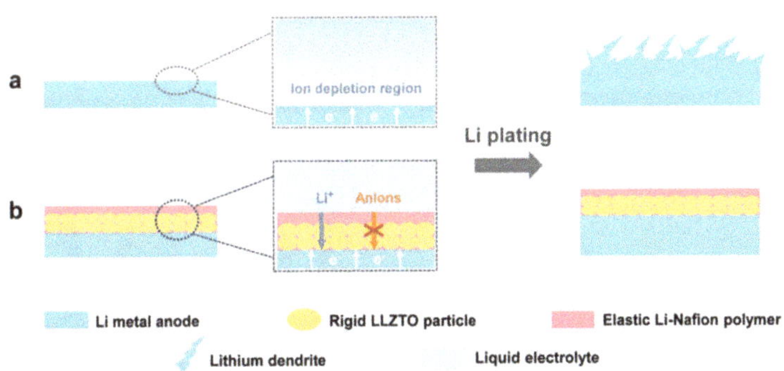

Figure 1.13. Schematic of organic/inorganic composite artificial SEI film induced uniform lithium deposition. Source: [81]. Reproduced with permission of Wiley-VCH Verlag GmbH & Co. KGaA, Weinheim.

1.2.3 Host framework composite

According to the aforementioned lithium deposition model, it is known that lithium nucleation plays a very important role in the plating and stripping process of metallic lithium, and the nucleation of lithium can be effectively controlled by designing the substrate or current collector. Among them, the use of a three-dimensional porous current collector instead of a two-dimensional planar current collector is one of the strategies to effectively inhibit the formation of lithium dendrites due to the following main features: (i) The high specific surface area can effectively reduce the local current density and achieve a uniform electric field distribution; (ii) the porous cross-linked structure can provide abundant lithium storage sites, and can alleviate the volume change caused by the process of lithium deposition and dissolution; (iii) the three-dimensional framework can promote the uniform nucleation and dense deposition of lithium [82–88].

1.2.3.1 Uniform electric field distribution

The three-dimensional current collector with porous structure can play the role of uniform electric field and charge distribution on the electrode surface. As shown in figure 1.14, in the early stage of formation of lithium dendrites, the lithium bumps formed on the planar current collector have a larger tip curvature, and the electric field strength near them is higher, which leads to preferential enrichment of lithium ions at the tips [89]. This uneven electric field distribution easily aggravates the growth of lithium dendrites. Compared with the planar current collector, the three-dimensional current collector can realize uniform electric field distribution by constructing rich charge centers, which is beneficial to the uniform deposition of lithium and the elimination of tip enrichment. In the case of three-dimensional porous copper current collector, the abundant convex structures on the three-dimensional porous structure serve as charge centers and nucleation sites during the lithium deposition process, which can make the electric field uniform and the negative charge along the copper conductive framework evenly distributed, leading the lithium ions to form a relatively flat deposition morphology.

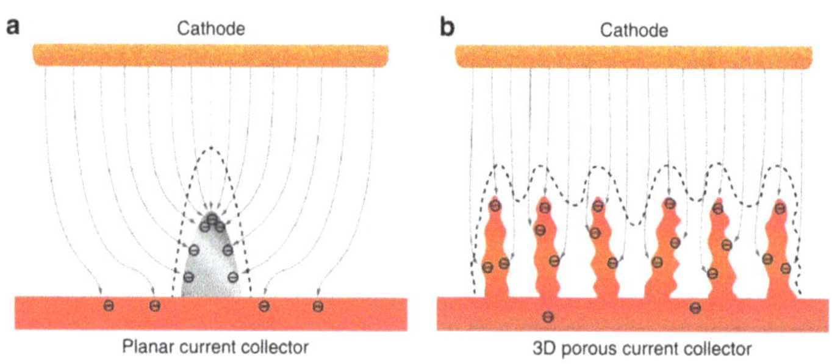

Figure 1.14. Schematic of different lithium deposition pattern on (a) planar current collector and (b) 3D porous current collector. Source: [89]. Reproduced with permission of Springer Nature.

1.2.3.2 The influence of Li^+ ion concentration gradient

During the lithium deposition process, the rapid consumption of Li^+ ions on the negative electrode surface will cause the electrolyte to form a Li^+ ion concentration gradient in the direction perpendicular to the electrode, resulting in concentration polarization. Therefore, the tip of the dendrite far away from the anode will have a higher concentration of Li^+ ions, which is easy to continue to grow. For the three-dimensional electrode structure, on the one hand, due to the decrease of local current density, the consumption rate of Li^+ ions per unit area of the three-dimensional negative electrode becomes slower, thereby alleviating the concentration polarization; on the other hand, the three-dimensional current collector with strong affinity or surface lithophilization modification can make it actively absorb lithium ions to uniformize the distribution of Li^+ ions and alleviate the tip effect, reducing the concentration gradient of Li^+ ions. For example, a layer of polyethyleneimine three-dimensional porous sponge (PPS) with lithophile functional groups is constructed on the surface of copper foil to induce uniform Li deposition [90]. Due to the presence of strong lithophile groups on the surface, PPS attracts positively charged Li^+ ions to form a stable interface layer with double charge (figure 1.15). Driven by the voltage, the Li^+ ions in the electric double layer migrate to the negatively charged electrode, which drives the movement of solvated molecules. At the same time, the friction force drives the electrolyte in the pores to flow directionally, thereby forming an electro-osmotic effect, which accelerates the mass transfer rate in the structure. In addition, PPS can enrich Li^+ ions in its sponge structure, making the Li^+ ion concentration higher than the external

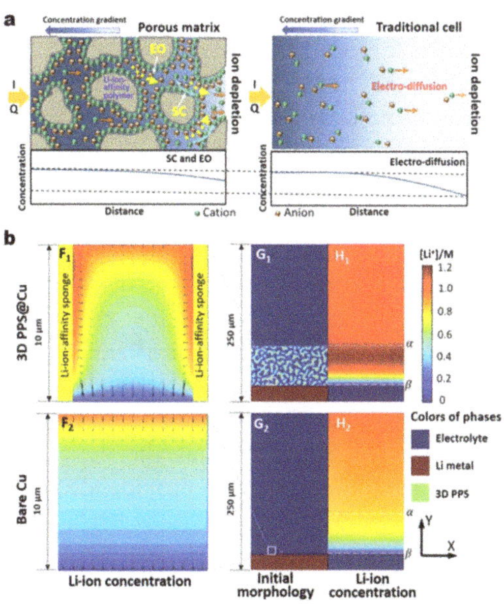

Figure 1.15. (a) Schematic of 3D PPS host and Li^+ ion electrokinetic self-concentrating pumping mechanism. (b) Equilibrium Li-ion concentration profiles with constant-reaction-current electrode surfaces for 3D PPS@Cu electrode and bare Cu electrode. Source: [90]. Reproduced with permission of Springer Nature.

electrolyte, effectively increasing the average Li^+ ion concentration during lithium deposition, weakening the negative impact of concentration polarization, and ultimately causing local current. The density distribution is consistent. Even at high current density (20 mA cm^{-2}), the uniform deposition of lithium metal in the pore structure can still be achieved.

1.3 Challenge of Li anode

In order to achieve high-energy density batteries, the academic and industrial circles have conducted more and more extensive research on lithium metal anodes. However, the performance degradation and safety hazards caused by the growth of lithium dendrites and unstable interface structures are also very prominent and seriously hinder the commercialization process for the lithium metal anode. At present, although researchers have proposed effective protection strategies for the lithium anode, and have made great progress, providing many new insights for understanding the interface reaction between electrolyte and metal electrodes, improving SEI, and constructing composite electrode materials. However, some basic and important issues regarding lithium metal anodes are still unresolved. On the one hand, some basic issues, such as uneven lithium deposition and interfacial reactions, have not been explored at the theoretical level, and the experimental observation results still need to be at the atomic level to reveal the underlying mechanism; on the other hand, although people have developed many effective methods to inhibit the growth of lithium dendrites, especially the introduction of new framework materials for Li deposition, the related theories and design principles have not been fully established.

(i) The mechanism of lithium nucleation and deposition needs to be further studied. During the Li^+ ion deposition process, the formation of dendrites is a spontaneous thermodynamic process. In order to control the surface morphology of lithium metal electrodes and inhibit the growth of lithium dendrites, the key factors affecting the kinetic process of lithium nucleation and the deposition morphology should be studied in depth. At present, based on diffusion thermodynamics and electrodeposition kinetics, Jana *et al* have calculated and simulated the deposition of Li^+ ions under different conditions, and proposed three growth mechanisms and six dendrite morphologies, which are consistent with the experiment and form a complete system [91]. Some new concepts, such as the proposed 'lithium bond', also affect the lithium-containing clusters and solvation structure in the electrolyte, as well as the deposition of lithium on the electrode surface, and are expected to provide ideas for the study of lithium nucleation mechanisms [92].

(ii) The deposition and dissolution behaviors of lithium metal anodes at different scales have corresponding mechanisms, and an in-depth understanding of the mechanism is inseparable from advanced characterization methods. Atomic-scale scientific problems involved with lithium ions, solvated lithium ions, lithium ions in the crystal lattice, and bulk lithium atoms, which is corresponding to the scanning electron microscope (SEM), atomic force microscope (AFM), and transmission electron microscope (TEM) on the atomic-scale behavior of the interface and bulk phase; at the micro-scale, it is mainly manifested as the growth behavior of the SEI film interface and initial nucleation, including electron energy loss spectroscopy, secondary ion mass spectrometry, selected area electron diffraction and nano x-ray tomography scanning characterization methods, which can be used to study the crystal structure, element composition, and microscopic morphology at the microscopic scale; at the mesoscopic scale, lithium metal presents a variety of morphologies. There are many research methods at this scale for detecting the information of the lower morphology and ion transport., including scanning electron microscopes, optical microscopes, and neutron depth spectroscopy, x-ray photoelectron spectroscopy, micron x-ray tomography, etc. Various characterization techniques for lithium metal anode in multiple spatial scales are summarized in figure 1.16 [93]. There are many kinds of lithium deposition behaviors on the surface of lithium metal anodes in solid–liquid systems, including lithium dendrites, dead lithium or macroscopic uneven deposition and dissolution. The influencing factors include external factors such as current density, charge and discharge capacity, temperature, pressure, as well as internal influencing factors such as large volume changes and high chemical reaction activity, ultimately leading to different failure modes of the battery. Therefore, it is necessary to systematically understand the mechanism behind macroscopic phenomena through multi-information dimensional and multi-scale research, and comprehensively design lithium metal anode protection strategies. The application of advanced experimental characterization technology can verify the calculation simulation results and inspire

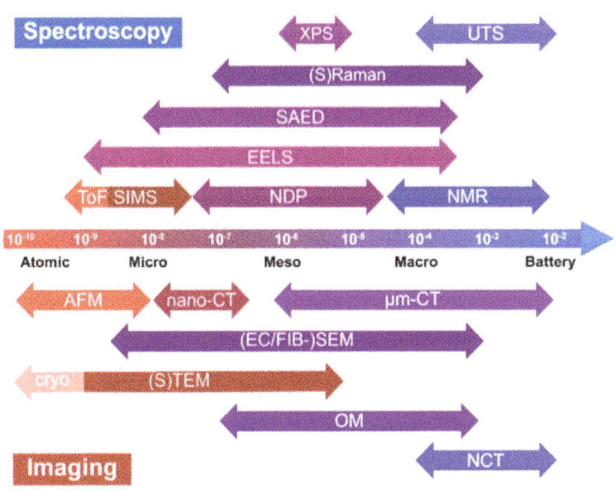

Figure 1.16. Characterization techniques for lithium metal anode in multiple spatial scales. Source: [93]. Reproduced with permission of China Association for Science and Technology.

new calculation simulation methods, which is an important way for in-depth study of lithium metal anodes in the future.

(iii) Compared with liquid batteries, lithium metal anodes have more practical potential in solid-state batteries. In liquid batteries, lithium metal anodes and electrolytes have serious side reactions, lithium dendrites formed during the cycle, and dead lithium. The safety and cyclability of the solid electrolyte have brought severe challenges. The solid electrolyte has higher mechanical strength, and it is possible to block the penetration of lithium dendrites from the macro-scale consideration, and inhibit the side reaction of lithium and the electrolyte. The extremely low electrode potential of the metal lithium negative electrode can increase the operating voltage of the battery. However, this also causes the solid electrolyte to be thermodynamically unstable to the metal lithium [7, 94–96]. Some solid electrolytes containing multivalent cations, such as lithium aluminum germanium phosphate (LAGP) and lithium titanium aluminum phosphate (LATP) with sodium ion fast ion conductor type (NASICON) structure, are thermodynamically unstable to lithium, which can be reduced and embedded in it to form an ion–electron mixed conductor. The formation of such interface can make Li^+ ions and electrons continue to penetrate, eventually causing self-discharge and even short-circuiting the battery. Among them, the tetravalent metal cations (Ge^{4+}, Ti^{4+}) are reduced to low-valent oxides or even metal elements. Besides, Ti^{4+} ions in the perovskite-type lithium lanthanum titanium oxide (LLTO) are also easily reduced to low-valent states oxides (Ti_2O_3 and TiO) and metallic titanium. Because of the formation of local electronic conduction paths, the interface layer will gradually become thicker. Even if the amount of these products is small, because the interface layer is not completely electrically insulating, the interface layer will continue to penetrate into the deep. In addition, the solid–solid contact between the lithium anode and the solid electrolyte will cause a huge interface impedance, and may promote the subsequent uneven reaction of the metal lithium, seriously deteriorating the electrochemical performance of the solid-state battery. The above issues will deteriorate the battery rate performance, coulomb efficiency and cycle stability, eventually causing the battery to fail. Therefore, in the follow-up research, a solid electrolyte that matches the lithium metal anode should be designed, and the physical contact and chemical stability of the solid–solid interface should be fully considered to improve the electrochemical performance of the LMBs.

References

[1] Li M, Lu J, Chen Z W and Amine K 2018 30 years of lithium-ion batteries *Adv. Mater.* **30** 1800561

[2] Aravindan V, Lee Y S and Madhavi S 2015 Research progress on negative electrodes for practical Li-ion batteries: beyond carbonaceous anodes *Adv. Energy Mater.* **5** 1402225

[3] Kim H, Jeong G, Kim Y U, Kim J H, Park C M and Sohn H J 2013 Metallic anodes for next generation secondary batteries *Chem. Soc. Rev.* **42** 9011–34
[4] Shen X, Liu H, Cheng X B, Yan C and Huang J Q 2018 Beyond lithium ion batteries: higher energy density battery systems based on lithium metal anodes *Energy Storage Mater.* **12** 161–75
[5] Lin D C, Liu Y Y and Cui Y 2017 Reviving the lithium metal anode for high-energy batteries *Nat. Nanotechnol.* **12** 194–206
[6] Zhang Y, Zuo T T, Popovic J, Lim K, Yin Y X, Maier J and Guo Y G 2020 Towards better Li metal anodes: challenges and strategies *Mater. Today* **33** 56–74
[7] Cheng X B, Zhao C Z, Yao Y X, Liu H and Zhang Q 2019 Recent advances in energy chemistry between solid-state electrolyte and safe lithium-metal anodes *Chem.* **5** 74–96
[8] Zhang X Y, Wang A X, Liu X J and Luo J Y 2019 Dendrites in lithium metal anodes: suppression, regulation, and elimination *Acc. Chem. Res.* **52** 3223–32
[9] Liu J *et al* 2019 Pathways for practical high-energy long-cycling lithium metal batteries *Nat. Energy* **4** 180–86
[10] Zhang X, Yang Y A and Zhou Z 2020 Towards practical lithium-metal anodes *Chem. Soc. Rev.* **49** 3040–71
[11] Cheng X B, Zhang R, Zhao C Z and Zhang Q 2017 Toward safe lithium metal anode in rechargeable batteries: a review *Chem. Rev.* **117** 10403–73
[12] Bruce P G, Freunberger S A, Hardwick L J and Tarascon J M 2012 Li-O-2 and Li-S batteries with high energy storage *Nat. Mater.* **11** 19–29
[13] Wang R H, Cui W S, Chu F L and Wu F X 2020 Lithium metal anodes: present and future *J. Energy Chem.* **48** 145–59
[14] Yasin G, Arif M, Mehtab T, Lu X, Yu D L, Muhammad N, Nazir M T and Song H H 2020 Understanding and suppression strategies toward stable Li metal anode for safe lithium batteries *Energy Storage Mater.* **25** 644–78
[15] Xu K 2014 Electrolytes and interphases in Li-ion batteries and beyond *Chem. Rev.* **114** 11503–618
[16] Liu B, Zhang J G and Xu W 2018 Advancing lithium metal batteries *Joule* **2** 833–45
[17] Yang C P, Fu K, Zhang Y, Hitz E and Hu L B 2017 Protected lithium-metal anodes in batteries: from liquid to solid *Adv. Mater.* **29** 1701169
[18] Lu D P *et al* 2015 Failure mechanism for fast-charged lithium metal batteries with liquid electrolytes *Adv. Energy Mater.* **5** 1400993
[19] Yue X Y, Ma C, Bao J, Yang S Y, Chen D, Wu X J and Zhou Y N 2021 Failure mechanisms of lithium metal anode and their advanced characterization technologies *Acta Phys. Chim. Sin.* **37** 2005012
[20] Gao N S J *et al* 2021 Fast Diagnosis of failure mechanisms and lifetime prediction of Li metal batteries *Small Methods* **5** 2000807
[21] Liu H, Cheng X B, Xu R, Zhang X Q, Yan C, Huang J Q and Zhang Q 2019 Plating/stripping behavior of actual lithium metal anode *Adv. Energy Mater.* **9** 1902254
[22] Chen X R, Yan C, Ding J F, Peng H J and Zhang Q 2021 New insights into 'dead lithium' during stripping in lithium metal batteries *J. Energy Chem.* **62** 289–94
[23] Zhu R J, Yang H J, Fadillah L, Xiong Z T, Kowalski D, Zhu C Y, Kitano S, Aoki Y and Habazaki H 2021 A lithiophilic carbon scroll as a Li metal host with low tortuosity design and 'Dead Li' self-cleaning capability *J. Mater. Chem.* A **9** 13332–43
[24] Jo H, Song D, Jeong Y C, Lee Y M and Ryou M H 2019 Study on dead-Li suppression mechanism of Li-hosting vapor-grown-carbon-nanofiber-based protective layer for Li metal anodes *J. Power Sources* **409** 132–38

[25] Qiu X G, Liu W, Liu J D, Li J Z, Zhang K and Cheng F Y 2021 Nucleation mechanism and substrate modification of lithium metal anode *Acta Phys. Chim. Sin.* **37** 2009012
[26] Chen X R, Zhao B C, Yan C and Zhang Q 2021 Review on Li deposition in working batteries: from nucleation to early growth *Adv. Mater.* **33** 2004128
[27] Guan X Z, Wang A X, Liu S, Li G J, Liang F, Yang Y W, Liu X J and Luo J Y 2018 Controlling nucleation in lithium metal anodes *Small* **14** 1801423
[28] Qiu X G, Liu W, Liu J D, Li J Z, Zhang K and Cheng F Y 2021 Nucleation mechanism and substrate modification of lithium metal anode *Acta Phys. Chim. Sin.* **37** 11
[29] Biswal P, Stalin S, Kludze A, Choudhury S and Archer L A 2019 Nucleation and early stage growth of Li electrodeposits *Nano Lett.* **19** 8191–200
[30] Li N, Ye Q, Zhang K, Yan H B, Shen C, Wei B Q and Xie K Y 2019 Normalized lithium growth from the nucleation stage for dendrite-free lithium metal anodes *Angew. Chem. Int. Ed.* **58** 18246–51
[31] Dong J, Dai H L, Fan Q F, Lai C and Zhang S Q 2019 Grain refining mechanisms: initial levelling stage during nucleation for high-stability lithium anodes *Nano Energy* **66** 104128
[32] Zhang R, Chen X R, Chen X, Cheng X B, Zhang X Q, Yan C and Zhang Q 2017 Lithiophilic sites in doped graphene guide uniform lithium nucleation for dendrite-free lithium metal anodes *Angew. Chem. Int. Ed.* **56** 7764–68
[33] Wei C L, Fei H F, An Y L, Tao Y, Feng J K and Qian Y T 2019 Uniform Li deposition by regulating the initial nucleation barrier via a simple liquid-metal coating for a dendrite-free Li metal anode *J. Mater. Chem.* A **7** 18861–70
[34] Pei A, Zheng G Y, Shi F F, Li Y Z and Cui Y 2017 Nanoscale nucleation and growth of electrodeposited lithium metal *Nano Lett.* **17** 1132–39
[35] Yan K, Wang J Y, Zhao S Q, Zhou D, Sun B, Cui Y and Wang G X 2019 Temperature-dependent nucleation and growth of dendrite-free lithium metal anodes *Angew. Chem. Int. Ed.* **58** 11364–68
[36] Huang Z J, Zhou G M, Lv W, Deng Y Q, Zhang Y B, Zhang C, Kang F Y and Yang Q H 2019 Seeding lithium seeds towards uniform lithium deposition for stable lithium metal anodes *Nano Energy* **61** 47–53
[37] Rehnlund D, Ihrfors C, Maibach J and Nyholm L 2018 Dendrite-free lithium electrode cycling via controlled nucleation in low LiPF6 concentration electrolytes *Mater. Today* **21** 1010–18
[38] Yan K, Lu Z D, Lee H W, Xiong F, Hsu P C, Li Y Z, Zhao J, Chu S and Cui Y 2016 Selective deposition and stable encapsulation of lithium through heterogeneous seeded growth *Nat. Energy* **1** 16010
[39] Liu Y, Li B, Liu J H, Li S M and Yang S B 2017 Pre-planted nucleation seeds for rechargeable metallic lithium anodes *J. Mater. Chem.* A **5** 18862–69
[40] Yi J S, Chen J H, Yang Z, Dai Y, Li W M, Cui J, Ciucci F, Lu Z H and Yang C L 2019 Facile patterning of laser-induced graphene with tailored li nucleation kinetics for stable lithium-metal batteries *Adv. Energy Mater.* **9** 1901796
[41] Yang C P, Yao Y G, He S M, Xie H, Hitz E and Hu L B 2017 Ultrafine silver nanoparticles for seeded lithium deposition toward stable lithium metal anode *Adv. Mater.* **29** 1702714
[42] Meng Q Q *et al* 2019 Heterogeneous nucleation and growth of electrodeposited lithium metal on the basal plane of single-layer graphene *Energy Storage Mater.* **16** 419–25
[43] Chazalviel 1990 Electrochemical aspects of the generation of ramified metallic electro-deposits *Phys. Rev. A: At. Mol. Opt. Phys.* **42** 7355–67

[44] Li H P, Ji X Y and Liang J J 2020 Dual-functional ion redistributor for dendrite-free lithium metal anodes *Rare Met.* **39** 861–62
[45] Huang G, Han J H, Zhang F, Wang Z Q, Kashani H, Watanabe K and Chen M W 2019 Lithiophilic 3D nanoporous nitrogen-doped graphene for dendrite-free and ultrahigh-rate lithium-metal anodes *Adv. Mater.* **31** 1805334
[46] Pan H *et al* 2021 Fast Li plating behavior probed by X-ray computed tomography *Nano Lett.* **21** 5254–61
[47] Bai P, Li J, Brushett F R and Bazant M Z 2016 Transition of lithium growth mechanisms in liquid electrolytes *Energy Environ. Sci.* **9** 3221–29
[48] Li S, Jiang M W, Xie Y, Xu H, Jia J Y and Li J 2018 Developing high-performance lithium metal anode in liquid electrolytes: challenges and progress *Adv. Mater.* **30** 1706375
[49] Pang Q, Liang X, Shyamsunder A and Nazar L F 2017 An *in vivo* formed solid electrolyte surface layer enables stable plating of Li metal *Joule* **1** 871–86
[50] Chang H J, Ilott A J, Trease N M, Mohammadi M, Jerschow A and Grey C P 2015 Correlating microstructural lithium metal growth with electrolyte salt depletion in lithium batteries using Li-7 MRI *J. Am. Chem. Soc.* **137** 15209–16
[51] Peng Z *et al* 2019 Enhanced stability of Li metal anodes by synergetic control of nucleation and the solid electrolyte interphase *Adv. Energy Mater.* **9** 1901764
[52] Hou C, Han J H, Liu P, Yang C C, Huang G, Fujita T, Hirata A and Chen M W 2019 Operando observations of SEI film evolution by mass-sensitive scanning transmission electron microscopy *Adv. Energy Mater.* **9** 1902675
[53] Cheng X B, Yan C, Chen X, Guan C, Huang J Q, Peng H J, Zhang R, Yang S T and Zhang Q 2017 Implantable solid electrolyte interphase in lithium-metal batteries *Chem.* **2** 258–70
[54] Wang D *et al* 2018 Zipper-inspired SEI film for remarkably enhancing the stability of li metal anode via nucleation barriers controlled weaving of lithium pits *Adv. Energy Mater.* **8** 1800650
[55] Li N W, Shi Y, Yin Y X, Zeng X X, Li J Y, Li C J, Wan L J, Wen R and Guo Y G 2018 A flexible solid electrolyte interphase layer for long-life lithium metal anodes *Angew. Chem. Int. Ed.* **57** 1505–9
[56] Zhai P B, Liu L X, Gu X K, Wang T S and Gong Y J 2020 Interface engineering for lithium metal anodes in liquid electrolyte *Adv. Energy Mater.* **10** 2001257
[57] Chen K H, Wood K N, Kazyak E, LePage W S, Davis A L, Sanchez A J and Dasgupta N P 2017 Dead lithium: mass transport effects on voltage, capacity, and failure of lithium metal anodes *J. Mater. Chem.* A **5** 11671–81
[58] Dai H L, Gu X X, Dong J, Wang C, Lai C and Sun S H 2020 Stabilizing lithium metal anode by octaphenyl polyoxyethylene-lithium complexation *Nat. Commun.* **11** 643
[59] Ma Y L *et al* 2018 Enabling reliable lithium metal batteries by a bifunctional anionic electrolyte additive *Energy Storage Mater.* **11** 197–204
[60] Eshetu G G, Judez X, Li C M, Bondarchuk O, Rodriguez-Martinez L M, Zhang H and Armand M 2017 Lithium azide as an electrolyte additive for all-solid-state lithium-sulfur batteries *Angew. Chem. Int. Ed.* **56** 15368–72
[61] Yan C, Yao Y X, Chen X, Cheng X B, Zhang X Q, Huang J Q and Zhang Q 2018 Lithium nitrate solvation chemistry in carbonate electrolyte sustains high-voltage lithium metal batteries *Angew. Chem. Int. Ed.* **57** 14055–59
[62] Li S, Luo Z, Li L, Hu J G, Zou G Q, Hou H S and Ji X B 2020 Recent progress on electrolyte additives for stable lithium metal anode *Energy Storage Mater.* **32** 306–19

[63] Zu C X, Azimi N, Zhang Z C and Manthiram A 2015 Insight into lithium-metal anodes in lithium-sulfur batteries with a fluorinated ether electrolyte *J. Mater. Chem.* A **3** 14864–70

[64] Qi S H, Wang H P, He J, Liu J D, Cui C Y, Wu M G, Li F, Feng Y Z and Ma J M 2021 Electrolytes enriched by potassium perfluorinated sulfonates for lithium metal batteries *Sci. Bull.* **66** 685–93

[65] Dai H L, Xi K, Liu X, Lai C and Zhang S Q 2018 Cationic surfactant-based electrolyte additives for uniform lithium deposition via lithiophobic repulsion mechanisms *J. Am. Chem. Soc.* **140** 17515–21

[66] Zhang X Q, Chen X, Cheng X B, Li B Q, Shen X, Yan C, Huang J Q and Zhang Q 2018 Highly stable lithium metal batteries enabled by regulating the solvation of lithium ions in nonaqueous electrolytes *Angew. Chem. Int. Ed.* **57** 5301–5

[67] Cui Y L, Liu S F, Wang D H, Wang X L, Xia X H, Gu C D and Tu J P 2021 A facile way to construct stable and ionic conductive lithium sulfide nanoparticles composed solid electrolyte interphase on Li metal anode *Adv. Funct. Mater.* **31** 2006380

[68] Deng K R, Han D M, Ren S, Wang S J, Xiao M and Meng Y Z 2019 Single-ion conducting artificial solid electrolyte interphase layers for dendrite-free and highly stable lithium metal anodes *J. Mater. Chem.* A **7** 13113–19

[69] Fan L S, Guo Z K, Zhang Y, Wu X, Zhao C Y, Sun X, Yang G Y, Feng Y J and Zhang N Q 2020 Stable artificial solid electrolyte interphase films for lithium metal anode via metal-organic frameworks cemented by polyvinyl alcohol *J. Mater. Chem.* A **8** 251–58

[70] Liu F F *et al* 2020 Lithium-ion conductive Li2S/Li2Se protection layer for stable lithium metal anode *Adv. Funct. Mater.* **30** 2001607

[71] Liu Y Y, Lin D C, Yuen P Y, Liu K, Xie J, Dauskardt R H and Cui Y 2017 An artificial solid electrolyte interphase with high Li-ion conductivity, mechanical strength, and flexibility for stable lithium metal anodes *Adv. Mater.* **29** 1605531

[72] Chen H *et al* 2019 Uniform high ionic conducting lithium sulfide protection layer for stable lithium metal anode *Adv. Energy Mater.* **9** 1900858

[73] Li N W, Yin Y X, Yang C P and Guo Y G 2016 An artificial solid electrolyte interphase layer for stable lithium metal anodes *Adv. Mater.* **28** 1853–58

[74] Han B *et al* 2020 Self-regulated phenomenon of inorganic artificial solid electrolyte interphase for lithium metal batteries *Nano Lett.* **20** 4029–37

[75] Liu W, Liu P C and Mitlin D 2020 Review of emerging concepts in SEI analysis and artificial SEI membranes for lithium, sodium, and potassium metal battery anodes *Adv. Energy Mater.* **10** 2002297

[76] Hu Z L, Zhang S, Dong S M, Li W J, Li H, Cui G L and Chen L Q 2017 Poly(ethyl alpha-cyanoacrylate)-based artificial solid electrolyte interphase layer for enhanced interface stability of Li metal anodes *Chem. Mater.* **29** 4682–89

[77] Lopez J, Pei A, Oh J Y, Wang G J N, Cui Y and Bao Z A 2018 Effects of polymer coatings on electrodeposited lithium metal *J. Am. Chem. Soc.* **140** 11735–44

[78] Wu C, Guo F H, Zhuang L, Ai X P, Zhong F P, Yang H X and Qian J F 2020 Mesoporous silica reinforced hybrid polymer artificial layer for high-energy and long-cycling lithium metal batteries *ACS Energy Lett.* **5** 1644–52

[79] Zhang Y J, Wang G Y, Tang L, Wu J J, Guo B K, Zhu M, Wu C, Dou S X and Wu M H 2019 Stable lithium metal anodes enabled by inorganic/organic double-layered alloy and polymer coating *J. Mater. Chem.* A **7** 25369–76

[80] Ye S F, Wang L F, Liu F F, Shi P C, Wang H Y, Wu X J and Yu Y 2020 g-C(3)N(4) Derivative artificial organic/inorganic composite solid electrolyte interphase layer for stable lithium metal anode *Adv. Energy Mater.* **10** 2002647

[81] Xu R, Xiao Y, Zhang R, Cheng X B, Zhao C Z, Zhang X Q, Yan C, Zhang Q and Huang J Q 2019 Dual-phase single-ion pathway interfaces for robust lithium metal in working batteries *Adv. Mater.* **31** 1808392

[82] Zhu M Q, Li B, Li S M, Du Z G, Gong Y J and Yang S B 2018 Dendrite-free metallic lithium in lithiophilic carbonized metal-organic frameworks *Adv. Energy Mater.* **8** 1703505

[83] Chen X, Chen X R, Hou T Z, Li B Q, Cheng X B, Zhang R and Zhang Q 2019 Lithiophilicity chemistry of heteroatom-doped carbon to guide uniform lithium nucleation in lithium metal anodes *Sci. Adv.* **5** eaau7728

[84] Wang H S *et al* 2019 An interconnected channel-like framework as host for lithium metal composite anodes *Adv. Energy Mater.* **9** 1802720

[85] Dong L, Nie L and Liu W 2020 Water-stable lithium metal anodes with ultrahigh-rate capability enabled by a hydrophobic graphene architecture *Adv. Mater.* **32** 1908494

[86] Fang R P, Chen K, Yin L C, Sun Z H, Li F and Cheng H M 2019 The regulating role of carbon nanotubes and graphene in lithium-ion and lithium-sulfur batteries *Adv. Mater.* **31** 1800863

[87] Zheng Y, Zheng S S, Xue H G and Pang H 2019 Metal-organic frameworks for lithium-sulfur batteries *J. Mater. Chem.* A **7** 3469–91

[88] Zheng Z J, Ye H and Guo Z P 2020 Recent progress in designing stable composite lithium anodes with improved wettability *Adv. Sci.* **7** 2002212

[89] Yang C P, Yin Y X, Zhang S F, Li N W and Guo Y G 2015 Accommodating lithium into 3D current collectors with a submicron skeleton towards long-life lithium metal anodes *Nat. Commun.* **6** 8058

[90] Li G X, Liu Z, Huang Q Q, Gao Y, Regula M, Wang D W, Chen L Q and Wang D H 2018 Stable metal battery anodes enabled by polyethylenimine sponge hosts by way of electrokinetic effects *Nat. Energy* **3** 1076–83

[91] Jana A, Woo S I, Vikrant K S N and Garcia R E 2019 Electrochemomechanics of lithium dendrite growth *Energy Environ. Sci.* **12** 3595–607

[92] Chen X, Bai Y K, Zhao C Z, Shen X and Zhang Q 2020 Lithium bonds in lithium batteries *Angew. Chem. Int. Ed.* **59** 11190–200

[93] Pan H Y, Li Q, Yu X Q and Li H 2021 Characterization techniques for lithium metal anodes at multiple spatial scales *Acta Phys. Chim. Sin.* **37** 2008091

[94] Krauskopf T, Richter F H, Zeier W G and Janek J 2020 Physicochemical concepts of the lithium metal anode in solid-state batteries *Chem. Rev.* **120** 7745–94

[95] Chen W *et al* 2018 Designing safe electrolyte systems for a high-stability lithium-sulfur battery *Adv. Energy Mater.* **8** 1702348

[96] Xiang J W, Yang L Y, Yuan L X, Yuan K, Zhang Y, Huang Y Y, Lin J, Pan F and Huang Y H 2019 Alkali-metal anodes: from lab to market *Joule* **3** 2334–63

IOP Publishing

Metal Electrodes for Battery Technologies

Yurong Ren and Jianmin Ma

Chapter 2

Metal anodes for metal batteries beyond Li batteries

Yong Liu, Feng Tao, Kaijia Feng, Han Qian, Xiaoliang Zhai, Huaihu Sun and Jianmin Ma

Due to the rapid development of portable electronic equipment, unmanned aerial vehicles, electric vehicles and other fields, there is an urgent need to develop high-energy-density batteries. Metal batteries are considered to be the most promising next-generation high-energy batteries due to their light weight and large capacity, especially lithium (Li), sodium (Na), potassium (K), magnesium (Mg), aluminum (Al), and zinc (Zn) metal batteries. However, due to the scarcity and uneven distribution of lithium metal resources, its cost is too high, while other metal batteries have relatively abundant reserves and low cost, so they have received widespread attention. In this chapter, we review the characteristics of Na, K, Mg, Al and Zn metal batteries and their development in recent years. At the same time, it is pointed out that related challenges and future research directions are of great significance to the development of metal batteries.

2.1 Na anodes for Na batteries

2.1.1 Na properties

Rechargeable sodium (Na) based batteries have attracted great interest due to sodium's high natural abundance, low cost and electrochemical similarities with lithium (Li) based batteries. Although Na and Li are neighbors in the alkali group with similar properties, the scientific challenges faced by the commercialization of the Li metal anode and Na metal anode are essentially different. Therefore, it is necessary to explore and understand respective electrochemical cycling behaviors of Li and Na in more depth. First, Na metal has low mechanical strength and is very soft because of its weak metal bond energy, it is easy to be cut by a knife. Na elements only exists in minerals, such as rock salt NaCl, due to its high chemical reactivity. The first ionization energy of Na atom is lesser, 495.8 kJ mol^{-1}, so its last

doi:10.1088/978-0-7503-4871-3ch2

electron is easily lost to form a Na^+ cation [1]. In detail, the chemical, physical and electrochemical properties of Na metal and Na ions are diverse from Li metal and Li ions in virtue of their different electronic structure and atomic proper characters, which are listed in table 2.1. Meanwhile, according to the current booming trend of the electronic equipment market, people's demand for Li will increase at a rate of 5%. It is now estimated that the existing exploitable Li sources can only support the current Li-ion battery market for perhaps 65 years in consideration of the lack of sustainability of Li-containing precursors (Li_2CO_3), and Li recovery can delay this available shortage to around 2100 [2]. As a result, the Li_2CO_3 price is \$6.5 kg^{-1} and will continue to rise [3]. By comparison, as the second element in the alkali metal group of the periodic table, Na not only has physic-chemical properties similar to Li, but also possesses fundamental superiority over Li in terms of raw material availability and price. Na-containing precursors have greater variety and quantity than Li all over the world and are widely distributed in the Earth's crust. For instance, sodium chloride (NaCl), sodium sulfate (Na_2SO_4), including sodium carbonate (Na_2CO_3), are naturally saved in Na deposit such as minerals, brine and seawater, among them, and the storage capacity of Na_2CO_3 in the United States alone reaches 23 billion tons, which led to the Na_2CO_3 price (\$0.5 kg^{-1}) being well below Li_2CO_3 [4]. Besides, Na offers another cost and weight preponderance that Li counterparts do not when one takes the collector material in batteries into consideration. Due to the alloying reaction between Li and Al at low potential, Al can merely be designated as the collector of the positive electrode in Li-ion batteries (LIBs) [5]. In other words, copper (Cu) must be employed as anode current collecting for Li-ion batteries, which will raise the difficult levels on the

Table 2.1. The difference of properties between Li and Na [4].

Properties	Parameter	Li	Na
Physical properties	Atomic weight (g mol^{-1})	6.9	23
	Atomic radius (Å)	1.52	1.86
	Melting point (K)	454	371
	Molar volume (cm^3 mol^{-1})	12.97	23.75
	Ionic radius (Å)	0.76	1.02
	Abundance in the Earth's crust (%)	0.0017	2.3
	Price (carbonate salt, \$ per ton)	17 000	390
Chemical properties	Electron configuration	[1, 2]	[1, 2, 8]
	1st ionization energy (kJ mol^{-1})	520.2	498.8
	Reactivity with Al	Yes	No
Electrochemical properties	Voltage vs. SHE (V)	−3.04	−2.71
	Desolvation energy with EC (kJ mol^{-1})	211.3	151.9
	Desolvation energy with PC (kJ mol^{-1})	218.0	157.3
	Desolvation energy with DEC (kJ mol^{-1})	207.7	147.5
Mechanical properties	Brinell hardness (MPa)	5	0.69
	Bulk modulus (GPa)	11	6.3

popularization of low price and portability of the battery. However, Na does not react with Al, reducing cost and weight by allowing both positive and negative collectors to apply Al foil [6].

2.1.2 Development of Na anode

Based on the above advantages, a Na-based energy storage system has ignited the enthusiasm of researchers. Among them, the sodium-ion battery (SIB) as the representative, which has the same working principle and battery structure as a LIB and was developed side-by-side with the LIB in the mid-1980s, having again received the attention of the scientific community in the past decade [7]. So far, significant progress has been made in the development of SIB cathodes, including but not limited to layered transition metal oxides ($NaFeO_2$, O3-Type $Na[Fe_{1/2}Mn_{1/2}]O_2$, P3-type Na_xCoO_2, *et al*), polyanionic compounds ($NaFePO_4$, $Na_2FeP_2O_7$, $Na_3V_2(PO_4)_3$, etc) and other various Na insertion materials, some of which show comparable properties to their LIBs counterparts [8]. For example, a high-performance metal oxide cathode (P2-type $Na_x[Fe_{0.5}Mn_{0.5}]O_2$) was recently discovered by Komaba *et al*, which shows an almost comparable gravimetric energy of ~500 Wh kg^{-1} (based on electrode), to the well-known Li analog $LiFePO_4$ [9]. Although their overall capacity is equal to or even higher than that of LIB graphite, the capacity is achieved in a wide voltage window, starting from ~1.5 V vs Na/Na^+ [10]. Compared with LIBs, the battery voltage it provides is generally low, which is the main reason for the lower energy of most SIBs. Therefore, it can be seen that the research process of cathode materials is relatively long, during this period, we should consider the selection of anode materials reasonably to further improve the capacity of sodium-based batteries. Yet, for SIB, the selection of anode materials cannot completely imitate the current anode materials of LIBs. For instance, the typical graphite is the most common anode material for LIBs. But it is not compatible with the Na^+ ion on account of its low electrochemical activity in the Na-ion system [11]. After comprehensive screening of anode materials, several types of promising Na anode materials were proposed, according to multiple approaches such as nanostructured, alloy and insertion reaction [12]. A number of leading reviews have discussed in detail the development process and challenges of SIB cathode and anode materials in recent years [13]. Numerous anode materials wherein one of the most foreground anode materials is Na metal, it is known as the 'Holy Grail' of the Na-based battery owing to high theoretical specific capacity (1166 mAh g^{-1}) and low electrode potential (−2.71 V vs. standard hydrogen electrode, SHE) [14]. Based on Na metal as an anode, a new concept of high energy battery, sodium metal batteries (NMBs), corresponding to lithium metal batteries (LMBs) has been put forward. Rechargeable NMBs with Na metal anodes and conventional insertion cathodes can achieve energy density comparable to the most advanced Li-ion batteries with graphite anodes. Nowadays, the Na metal anode is widely used in many types of NMBs, besides ordinary Na metal batteries, it also includes $Na–O_2$, $Na–CO_2$, $Na–SO_2$, and high-temperature/room-temperature Na–S (HT/RT Na/S) batteries etc, which derive from the cathode candidate, greenhouse gas (CO_2), extensive oxygen (O_2), sulfur (S), or sulfur dioxide (SO_2) and hold broad prospects thanks to ultra-high theoretical energy densities as shown in figure 2.1 [2] In this section, we will focus on the recent works with the Na metal anode and their practical application in NMB systems, including Na–O_2, Na–CO_2 and Na–S batteries, as well as discuss the influence of the main components of these batteries.

The metal–O_2 battery is one of the limited battery systems with a metal anode, which roughly compete with gasoline with respect to theoretical energy density. In accordance with only the weight of the negative material, the theoretical specific energy of Li–O_2 and Na–O_2 batteries can reach 11 430 and 3451 Wh kg^{-1}, respectively, which are the highest amongst all kinds of metal–O_2 battery systems [15]. In 2011, Peled *et al* reported a battery system operating at temperatures above 98 °C to replace Li–O_2 batteries, which indicates that the research on the Na–O_2 battery is rising [16]. Then in 2012, Sun *et al* followed up their initial work and launched the first Na–O_2 battery that can be recharged at room temperature [17]. The electrochemical reaction during the discharge cycle of the Na–O_2 battery can be expressed by the following chemical equations [15]

$$\text{Anode: } Na \rightarrow Na^+ + e^- \tag{2.1}$$

$$\text{Cathode: } O_2 + e^- \rightarrow O_2^- \tag{2.2}$$

$$\text{Equivalent to} \quad O_2 + 2e^- \rightarrow O_2^{2-} \tag{2.3}$$

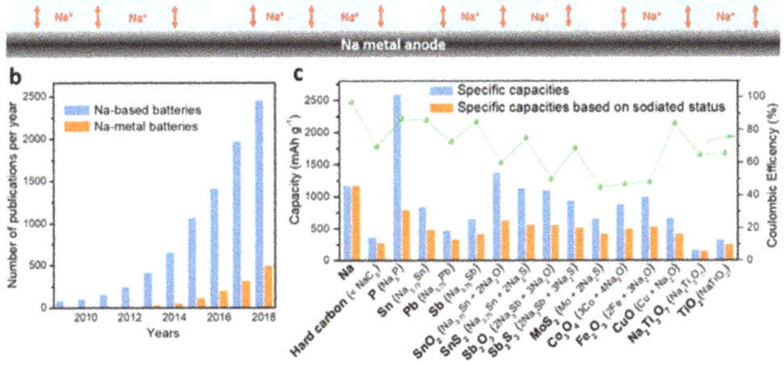

Figure 2.1. (a) The development schedule of the sodium-based battery and the schematic diagram and working principle of four typical rechargeable sodium metal batteries. (b) The number of scientific research articles on sodium-based battery and sodium metal battery published every year. The data came from 'Web of Science' with the keywords of 'sodium battery' and 'sodium metal anode'. (c) Theoretical specific gravity capacity and first cycle coulomb efficiency of various reported anode materials for room temperature (RT) sodium-ion batteries. Source: [15]. Reproduced with permission of John Wiley and Sons.

$$\text{Total: } Na + O_2 \rightarrow NaO_2 \quad E^0 = 2.27\ V \tag{2.4}$$

$$\text{Equivalent to} \quad 2Na + O_2 \rightarrow Na_2O_2 \quad E^0 = 2.33\ V \tag{2.5}$$

As mentioned above, the Na–O_2 battery has attracted more and more attention as a substitute for LIBs due to its high theoretical energy density. In addition, compared with the electrochemical semi-irreversible oxygen/superoxide (O_2/O_2^-) redox pair in the Li–O_2 system, the Na–O_2 battery enjoys advantages of high energy efficiency, which can be attributed to the reversible electrochemistry of the oxygen/superoxide (O_2/O_2^-) redox pair [18]. Despite the Na–O_2 battery being provided with the above merits, there are still some basic problems that are hard to solve, for example, the method of determining the key experimental parameters controlling the reaction pathway is still unclear. More than this, on the cathode side, the O_2/O_2^- generated from the cathode may not only migrate to the separator and metal Na through the concentration gradient of electrolyte, but also reacts with the Na metal to form an insulating oxide layer on the surface of Na, which prevents the transport of Na ions.

Of course, one of the most cardinal constraints is dendrite growth as usual, which is the same as that of LMBs on the anode side [19]. The formation of dendritic Na in Na–O_2 cells was demonstrated in Bi *et al*'s paper, and more than 80% of the glass fiber separators were penetrated by dendrites [20]. Another reason for the failure of the Na–O_2 battery is the high activity reaction between a Na metal anode and active oxygen in electrolyte. As a result, severe electrolyte decomposition and subsequent Na metal consumption occurs [21]. In view of the above predicament, researchers launched a series of measures to elevate the stability of the Na–O_2 battery. Firstly, people adopt the carbon to air electrode, for one thing it has good electrochemical stability and conductivity, for another thing it possess easy processability [15]. Furthermore, Sun *et al* reported the specific discharge capacity of the Na–O_2 battery is positively correlated with the surface area of the air electrode material [22]. A three-dimensional air electrode with two pore system constructed by nitrogen-doped carbon nanotubes vertically arranged on carbon paper was prepared by Yadegari and co-workers to resolve the mass-transfer limitations in the air electrode [23]. Secondly, it is found that the component part of the electrolyte has a decisive effect on the discharge products, electrolyte stability and composition of solid electrolyte interface (SEI) after study. At present, lithium formate (HCO_2Li), lithium acetate (CH_3CO_2Li), lithium propyl-dicarbonate ($C_3H_6(CO_2Li)_2$), lithium carbonate (Li_2CO_3), and water (H_2O), which have been observed in the Li–O_2 system [24]. These by-products directly or indirectly endanger the cycle life of batteries. At present, a large number of researchers use glycol dimethyl ether based electrolytes for avoiding parasitic reactions on account of it is compatible with Na and superoxide radicals [25].

Certainly, the side reaction also involves the negative electrode of Na metal, thus people put a generous energy into the anode side too, which is not only for coping with parasitic reaction but other fundamental obstacles that are fragile in SEI formation, irresistible dendrite growth, and violent volume change during charging/discharging. On the one side, due to the low reduction potential of the Na metal, compared with organic solvents, when the fresh Na metal is exposed in the electrolyte, it will quickly react with the electrolyte and be oxidized [26]. At the same time, the solvent and Na salt are reduced, forming a SEI, an electric insulator yet an ionic conductor, on the surface of the electrode to protect the Na metal [27].

A great amount of research has shown that the constituent of the passivation layer (SEI) at the interface between the Na electrode and electrolyte has a profound and lasting influence on the electrochemical behavior of the battery. Lutz and colleagues revealed that anions of Na salt in electrolyte containing-DME play an essential role in SEI formation of the Na–O_2 battery [28]. Similar to some reports of Li metal batteries, the formation of metal fluoride salts in SEI, such as NaF, seems to be the crucial component to enhance the stability of the Na anode in Na–O_2 batteries. It comes against the excessive parasitic reactions, limiting the O_2 crossover, rendering a low surface energy barrier with fast ion diffusion rate, and better adapting to volume changes [29]. Then, a serviceable protected layer consisting of NaF was designed by Zhou *et al* on the surface of the Na anode to realize a stable Na–O_2 battery [30].

Not just for sodium oxygen batteries, many researchers also have developed a series of methods to reform other types of Na-based battery. Recently, Wang's team studied the modification of surface film on the Na negative electrode by sodium polysulfide (Na_2S_6) in ether electrolyte. It was found that SEI was mainly composed of Na_2O, Na_2S_2 and Na_2S, which is beneficial to the circulation of a Na anode under high deposition capacity of 5 mAh cm^{-2} [31]. Subsequently, Wang *et al* proposed a bifunctional additive in tetraethylene glycol dimethyl ether (TEGDME) electrolyte, potassium bis(trifluoromethylsulfonyl)imide (KTFSI), based on the electrostatic shielding effect and the existence of $TFSI^-$ anions, which is conducive to constructing nitrided SEI layers, also enhancing the lifetime of Na metal anodes [32].

Meanwhile, the quality of Na^+ permeable but flexible were possessed by artificial SEI, which has evoked extensive discussion among peers [33]. At present, artificial SEI is mainly constructed on Na metal through the following strategies, including chemical pretreatment, physical technology and independent protective film. In addition to NaF, Na_3N and other inorganic sodium salts, some organic salts, such as sodium phenyldithiol (PhS_2Na_2), are profitable components in the protective film for planar Na electrodeposition too [34]. Afterwards, the physical coating on a metal Na anode has been proposed and widely studied since it can effectively passivate metal Na. For example, atomic layer deposition (ALD) technology is used to obtain a very thin protective film, including ALD–Al_2O_3 coating [35].

Apart from these methods, there are concentrated electrolyte, alloying methods to overcome the inherent shortcomings of Na metal [36]. Wang *et al* reported a double-layer SEI for a Na metal anode [37]. The upper layer is a composite of organic polymer and NaF, and the bottom layer is Na_3Sb alloy. Through the liquid based reaction between SbF_3 and Na ($3Na + SbF_3 \rightarrow 3NaF + Sb$; $Sb + 3Na \rightarrow Na_3Sb$), Na was designed to be immersed in diluted SbF_3 DMC solution to modify the structure of SEI. XPS analysis at different depths confirmed the mechanism of Na modification by the Na alloy fluoride grading layer, which can provide excellent cycle stability in carbonate and ether based electrolytes. Yao and co-workers deposited a 20 nm-thick Sn layer on Na metal inside a thermal evaporator equipped with a thickness monitor, which provides a 'natrophilic' decorative layer named $Na_{15}Sn_4$ to improve the wettability of the Na metal anode on electrolyte, thus improving the life of the Na symmetrical battery [38].

Salt concentration have a non-negligible effect on the properties of electrolyte, the stability of electrode/electrolyte interface and dendrite growth in virtue of the increase of initial metal ion concentration in electrolyte, the time of sand will also increase, thus delaying the formation of Na dendrites. After years of research, high concentration electrolyte and local high concentration electrolyte formed by adding diluent have been proposed in conjunction with being applied to Na-based batteries [39]. The above is the strategy of electrolyte design to protect the Na metal.

Next, the nanostructured collector and host are prepared to realize uniform deposition and stripping of the Na metal and reduce volume change during cycling [40]. In terms of collector, Jiang *et al* reported that three-dimensional (3D) porous Cu can significantly improve the stability of Na electroplating/stripping. After 400 cycles at 1 mA cm^{-2} and low voltage hysteresis voltage (20 mV), the CE value is as high as 99.4% [41]. In a manner of host, it should show good 'sodium affinity' to promote Na metal injection, high surface area to reduce local current density, and good stability to avoid side reaction with the electrolyte. Up to now, several main hosts have been exploited for a Na-based composite anode, and their electrochemical performance has been significantly heightened [42].

In the development process of the sodium air battery, besides the sodium oxygen battery, there is also the sodium carbon dioxide battery (Na–CO_2 battery). As we know, CO_2 is the main greenhouse gas, and its increasing emissions will lead to the disastrous consequences of climate change, sea level rise and glacier melting [43]. Therefore, in order to turn waste into treasure, scholars all over the world are committed to developing various chemical and physical ways to capture and store CO_2 or convert CO_2 into value-added materials. Among many energy storage technologies, the metal–CO_2 battery, a kind of metal air battery, has become one of the most attractive choices because of its unique characteristics of consuming greenhouse gas CO_2 and generating electric energy at the same time. Therefore, magnanimous manpower and material resources have been investment in this field to study its principle and application value. Thanks to high energy density (1125 Wh kg^{-1}), high degree of similarity with the corresponding Li–CO_2 battery, efficient utilization of CO_2 and high working voltage (2.0 V) of the Na–CO_2 battery (vs. Mg and Al, <1.0 V), it has attracted much attention [44]. In general, when assembling a Na–CO_2 battery, Na is used as the negative electrode, CO_2-involved porous material with enough gas diffusion path and effective electrocatalyst is used as the cathode, and finally an ion conductive separator immersed in Na salt is assembled between the positive and negative electrodes. The working voltage of a Na–CO_2 battery is 2.35 V (relative to Na/Na^+) [45]. The charge/discharge mechanism is expressed by equations [46]:

$$\text{Anode: } 4Na \leftrightarrow 4Na^+ + 4e^- \tag{2.6}$$

$$\text{Cathode: } 4Na + 3CO_2 + 4e^- \leftrightarrow 2Na_2CO_3 + C \tag{2.7}$$

Although the Na–CO_2 battery has the above merits, there are still some essential problems in its various components, such as uncontrollable growth of Na dendrite caused by non-uniform deposition and exfoliation of Na metal, continuous parasitic reaction between Na metal and electrolyte, and the sluggish kinetics of the CO_2 reduction reaction (CRR)/CO_2 evolution reaction (CER) procedure during the repeated charge/discharge process [47]. In the aspect of a Na anode protection, Chen *et al* imitated the combination of Li metal and reduced graphene oxide (rGO), and also fabricated the corresponding composite anode of reduced graphene oxide and Na (rGO@Na) to elevate the lifetime of the Na–CO_2 battery [42]. After that, Luo *et al* quantitatively analyzed the addition amount of rGO, and deeply investigated the reaction activity between a rGO@Na anode and electrolyte, as well as the mechanical properties such as the strength and hardness of the electrode itself [42]. More importantly, in order to promote the reduction and precipitation of carbon dioxide, efficient catalysts are urgently needed. Since the introduction of rechargeable Na–CO_2 batteries, carbon nanotubes have been mainly used as cathode catalysts, which has excellent conductivity, large surface area, porous 3D network and rich channels. Chen's group successfully created multiwall carbon nanotubes (MWCNTs) as the positive electrode of the high-temperature all-solid-state Na–CO_2 battery [48]. At a cut-off capacity of 500 mAh g^{-1}, the all-solid-state flexible Na–CO_2 battery shows a low discharge charge voltage difference of 0.6 V. Even after 240 cycles, the overpotential is still as low as 1.0 V. This stable cycling performance is closely related to the high catalytic activity of MWCNTs for reversible conversion of discharge products. Furthermore, the wettability of the CO_2/cathode/electrolyte interface can be improved by some organic solvents, such as TEGDME, because it has penetrated into the inner wall of the active MWCNT, which is revealed by characterization of Raman and IR (infrared). Except for carbon-based catalysts, there are transition and noble metal oxide catalysts that have also been shown to have good electro-catalytic activity, like a bimetallic oxides Co_2MnO_x nanowire-decorated carbon fibers (CMO@CF) and ruthenium nanoparticles (Ru@KB) [49].

A metal sulfur battery is an attractive choice because of its abundant sulfur cathode, which can provide a very high theoretical capacity of 1672 mAh g^{-1} after full discharge [50]. A sodium sulfur battery (Na–S) is formed by combining sulfur (S) and the same rich Na as raw materials in the energy storage environment. At present, there are three types of Na–S battery derived in different temperature ranges: high temperature, (intermediate temperature) intermediate temperature and room temperature (RT) [51]. High temperature Na–S batteries were previously developed and operated at 300 °C. In this system, the two electrodes including S positive electrode and Na negative electrode are in the molten state, which leads to low energy efficiency, corrosion and safety risks. In addition, lowering the operating temperature of the battery will result in lower material costs associated with low-cost polymer sealing and battery packaging, and reduce thermal management issues, such as heat loss that can increase cycle life. To overcome these problems, a RT Na–S battery was first demonstrated in the 2006 [52]. In this part, we will focus on the development and challenge of the RT Na–S battery. The discharge reaction between a S cathode and Na anode at different voltage-plateau can be expressed as follows [52]:

When the voltage platform is about ~2.2 V, corresponding to a solid−liquid transition from elemental S to polysulfide of Na_2S_8:

$$S_8 + 2Na + 2e^- \rightarrow Na_2S_8 \tag{2.8}$$

When the voltage platform is at 2.20–1.65 V, corresponding to a liquid–liquid reaction:

$$Na_2S_8 + 2Na^+ + 2e^- \rightarrow 2Na_2S_4 \tag{2.9}$$

A low voltage-plateau is about 1.65 V, corresponding to a liquid–solid transition:

$$Na_2S_4 + 2/3Na^+ + 2/3e^- \rightarrow 4/3Na_2S_3 \tag{2.10}$$

$$Na_2S_4 + 2Na^+ + 2e^- \rightarrow 2Na_2S_2 \tag{2.11}$$

$$Na_2S_4 + 6Na^+ + 6e^- \rightarrow 4Na_2S \tag{2.12}$$

When the voltage platform at 1.65–1.20 V, corresponding to a solid–solid reaction:

$$Na_2S_2 + 2Na^+ + 2e^- \rightarrow 2Na_2S \tag{2.13}$$

The capacity and discharge voltage is about 1.65 V depend on the competition of species in equations (2.10)–(2.12). Due to the insulating properties of Na_2S_2 and Na_2S, the kinetics of voltage platform at 1.65–1.20 V is slow.

A RT Na–S battery follows a similar working principle as the Li–S battery, and has many challenges facing similar battery system. The main challenges are as follows: first, the insulation of Na_2S discharge products; the second is the dissolution of polysulfides; the third is the shuttle effect of polysulfide and dendrite formation on Na [50, 53]. In order to investigate and overcome the problems of the RT Na–S battery, scientists have focused mostly on the development of functional nano-composites, using electrolyte with excellent function and the construction of new battery configurations.

On the anode side, sodium atoms are 85% larger than lithium atoms, and ions may be 25% or 55% larger depending on coordination, which leads to an interesting phenomenon, that is, when operating at room temperature, Na are more prone to uneven deposition/stripping than Li to form dendrites and unstable deposits [54]. This will further cause the formation of inhomogeneous SEI films. A Na–Sn–C alloy anode was reported by Lee *et al* to enhance the capacity retention, but this brings the side effect of reducing the output voltage of the sodium sulfur battery [55]. Cui *et al* developed a highly reversible and non-dendritic Na metal anode which can be stripped by sodium hexafluorophosphate ($NaPF_6$) in glymes (mono-, di- and tetraglyme) without any additional modification measures [56]. This outstanding performance is mainly contributed by SEI rich in Na_2O and NaF.

From the cathode point of view, the preparation of flexible current collector combined with rigid active materials is a widely used method, such as hydrothermal reaction, photothermal reduction, pulsed laser deposition, ultrasonic and co-deposition, casting and drying [7, 57]. Further, a method to avoid the combination of collector and active material was adopted by Yu and colleagues [58]. That is employing a binder-free, self-weaving MWCNT fabric electrode. It is easy to study the electrochemical characteristics of the low pressure platform region without considering any influence of the conversion process of elemental sulfur to long-chain sodium polysulfide. Not just MWCNTs, a variety of carbon materials (1D, 2D and 3D) have been used as conductive skeletons in S/C composite electrodes [59]. Among them, mesoporous carbon is the most attractive because their pores in the matrix can be doped with sulfur and provide conductivity and suitable space. The volume changes during the operation of the battery (up to 74%).

2.1.3 Challenge of Na anode

Although some progress has been made in the development of a Na metal anode, there are still major challenges to be overcome, especially in the practical application of NMBs (figure 2.2). First and foremost, although the physicochemical properties of Na metal are similar to that of Li metal, Li metal has been studied for decades and made great progress, especially in dendrite growth and protection of Li metal. In contrast, the RT Na metal battery system is still in its infancy and has great development space. To be specific, in terms of cycle life, Li metal symmetric batteries can show obvious voltage polarization after thousands of cycles, while for Na metal symmetric batteries, the cycle life is one order of magnitude different, that is, within hundreds of hours [60]. Then, in terms of large current density and high capacity, the Li metal anode has achieved stable cycling at 10 mA cm^{-2} and 10 mAh cm^{-2} [60]. Yet, the capacity of 3 mAh cm^{-2} and the current density of 5 mA cm^{-2} are the ultimate limit of the current of Na metal batteries, which is far from meeting the threshold of practical application. Furthermore, different from the detailed study of

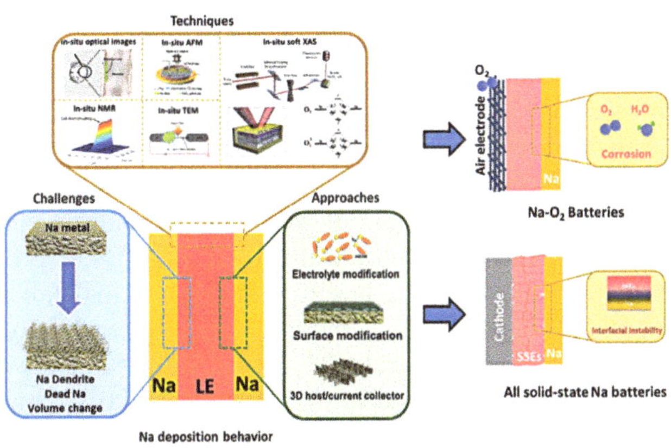

Figure 2.2. The deficiency of the development of a Na metal anode and the protection strategy of its practical application. Source: [19]. Reproduced with permission of The Royal Society of Chemistry.

the deposition/stripping of Li metal, few studies have involved a basic understanding of the stripping/plating of a Na metal anode, including the structure of Na deposition, like moss and dendrite Na. In particular, the relationship between Na deposition and other parameters, such as current density, capacity, temperature and pressure. In addition, the nucleation process of Na and the mechanism of SEI film formation should be further discussed, and more work should include the study of nucleation location, time and the composition along with regulation of SEI film. Finally, despite a variety of characterization techniques having been used to study sodium anodes, the correctness of sodium nucleation, dendrite growth and chemical/electrochemical reaction mechanism is still unclear, and more advanced techniques and characterization are needed. In the study of the SEI layer on Na metal surface, the synchrotron radiation will be one of the excellent candidate technologies, situ technology in particular. In addition, x-ray techniques of synchrotron, such as x-ray absorption spectroscopy (XAS) and XPS, are powerful tools for understanding SEI formation on Na metal surface. Moreover, x-ray tomography microscope with micron spatial resolution can be used for 2D and 3D study of sodium deposition processes and dendrite formation. The transmission x-ray microscopy can simultaneously feedback the structure and chemical information of the samples on the nanoscale (as low as 20 nm), which is helpful for the study of the growth of Na dendrites and SEI composition. At the same time, due to the high reactivity of Na metal, the most authentic chemical information can be obtained without changing the composition of Na dendrite or SEI layer. These characterization methods are more or less helpful for us to understand the chemical composition of sodium metal surface and SEI film, but they are not comprehensive and in-depth.

2.2 K anode for K batteries

2.2.1 K properties

Potassium (K) is the next alkali metal element in the periodic table after sodium. It has lower reduction potential than sodium, which makes KIB operate at higher potential (K: −2.93 V and Na: −2.71 V vs. SHE) [61]. This potential difference may result in potassium-based batteries with higher energy density than sodium-based batteries. Besides, potassium is abundant in the crust, ranking seventh in reserves, and the cost of potassium carbonate is much lower than that of lithium carbonate. In addition, potassium is extremely sensitive to oxygen and water in the air, leading to very rapid oxidation or hydration. Another feature similar to sodium is that the anode part of a potassium-based battery does not need a copper current collector, because potassium and aluminum do not form an alloy [62]. A series of properties such as physics and electrochemistry for Li, Na and K are compared in detail in table 2.2 [63].

In the light of the data in table 2.2, the property difference between K and Li including Na becomes obvious immediately. The lattice parameters of K are 34.1% and 23.3% larger than those of Na and Li (0.533, 0.429 and 0.351 nm), respectively, which leads to more expansion/contraction of the metal film per plating/stripping. Only a 64 °C melting point is the lowest among the three alkali metals, indicating that K is prone to volume and surface self-diffusion at room temperature, leading to

Table 2.2. Comparison of various physical and electrochemical properties of K, Na and Li.

		K	Na	Li
Element properties	Atomic number	19	11	3
	Atomic mass (g mol^{-1})	39.10	23.00	6.94
	Crystal structure/lattice parameter (Å)	Bcc, Im3m/ 5.33	Bcc, Im3m/ 4.29	Bcc, Im3m/ 3.51
Physical properties	Melting temperature (°C)	64	98	181
	Shear modulus (GPa)	1.3	3.3	4.2
	Brinell hardness (MPa)	0.36	0.69	5
Conductivity properties	Desolvation energy in EC (kJ mol^{-1})	114.6	152.8	208.9
	Desolvation energy in PC (kJ mol^{-1})	119.2	158.2	215.8
	Ionic conductivity of electrolyte of 1 M MFSI in EC-DEC (mS cm^{-1})	10.7	9.7	9.3
	Ionic conductivity of electrolyte of 1 M MPF_6 in EC/DMC (mS cm^{-1})	11.5	10	9.8
Electrochemical properties	Potential in PC vs Li+PC/Li (V)	−0.09	0.23	0
	Gravimetric specific capacity of metal electrode (mAh g^{-1})	685	1166	3860
	Volumetric specific capacity (mAh cm^{-3})	589	1131	2062

rapid creep deformation under tensile and shear stress. Potassium has an order of magnitude lower hardness and four times lower shear modulus than lithium, which indicates that potassium is much better than lithium in the particular area of plastic flow and compliance. According to the established mechanical blocking theory (e.g. Monroe and Newman), this should greatly reduce the tendency of potassium dendrite to pierce the SEI and cell separator. In fact, the reality is that the growth of potassium dendrites and the associated electrical short circuit are much more serious in ester and ether electrolyte solvents. Moreover, it has been proved in previous reports that the diffusion rate of K^+ ion is equal to or even faster than that of Li^+ ion, which means that the classical ion concentration polarization localized field model of Sand's time cannot fully explain the unstable K metal interface [64].

2.2.2 Development of K anode

In view of the above properties of potassium, researchers have sprung up the exploration of the energy system for potassium storage. At present, intercalation chemistry based on lithium ion has successfully launched commercial LIB from portable devices to large-scale energy storage systems. Like lithium ion, potassium ions could insert/extract reversibly between positive electrode and negative electrode during repeated charge and discharge processes, and the generated energy could be applied for drive equipment. So, a series of energy storage technologies represented by potassium-ion battery (KIB), working principle and battery structure are the same as those of LIB, have become another possible energy storage system. To this day, significant progress has been made in the development of KIB anodes, including but not limited to graphite, various types of carbonaceous, metals and alloys, intercalation compounds, organic compounds, conversion-type anode materials and other potassium intercalation materials, some of which exhibit properties comparable to their LIB counterparts. For instance, Komaba *et al* carefully analyzed the binder in electrode materials and found that when sodium polyacrylate (Pana) was used as binder instead of polyvinylidene fluoride (PVDF), the rate performance, cycle stability and coulomb efficiency of the first cycle could be significantly improved [63]. Furthermore, several other forms of carbonaceous have been successfully fabricated and investigated as anode materials for KIBs [65]. For instance, a nitrogen-doped carbon microsphere was reported by Chen and co-workers, which exhibited excellent rate capability and long-term cycling stability [65]. In the aspect of intercalation anode, based on the widely studied titanate-based compounds used in LIBs and SIBs, Kishore *et al* and Han *et al* suggested the use of titanate-based, such as $K_2Ti_4O_9$ and $K_2Ti_8O_{17}$, whose layered structures are suitable for carrying potassium ions [66]. Beyond these inorganic electrodes, organic electrodes have been gradually developed in the vicinity of K-based batteries. Deng *et al* and Lei *et al* recently reported K_2TP and K_2PC, which build-up on a layered structure with K^+ transmission channels [67]. Apart the above anode materials, potassium metal began to grow up as a functional anode in potassium metal batteries (KMBs) and as a counter electrode in half cell KIBs. The reversible capacity of potassium (theoretical 687 mAh g^{-1}) is almost twice that of graphite containing Li^+ ion, which is much higher than that of most available K-ion anodes [68]. Combined with the lowest anode voltage in KMB, K metal is very attractive to energy devices in principle. It is also exciting to understand the electrochemical behavior of potassium metal anode, which provides a new palette for exploring various interface phenomena. But, to date, potassium metal anodes are neither well-controlled nor well-understood. As time goes on, K–metal anodes are widely used in various types of KMBs, including K–Se, K–O_2 and K–S batteries in addition to ordinary KMBs, which are derived from cathode candidates, non-metallic elements (Se), large amounts of oxygen (O_2) and sulfur (S), as you can see in figure 2.3. In this portion, we will focus on the recent work of a K metal anode and its practical application in KMB system, including K–S, K–O_2 and K–Se batteries, and discuss the effect of the composition of these batteries on them.

At the same time, as one of the substitutes of LIB, the RT K–S battery has received considerable attention because of its low-cost and huge energy-density [69]. The S cathode of a K–S battery can provide high theoretical capacity (1675 mAh g^{-1}), which is ~10 times that of conventional LIB [70]. Not only in terms of capacity, but also in terms of storage abundance, low price and environmental friendliness further promote

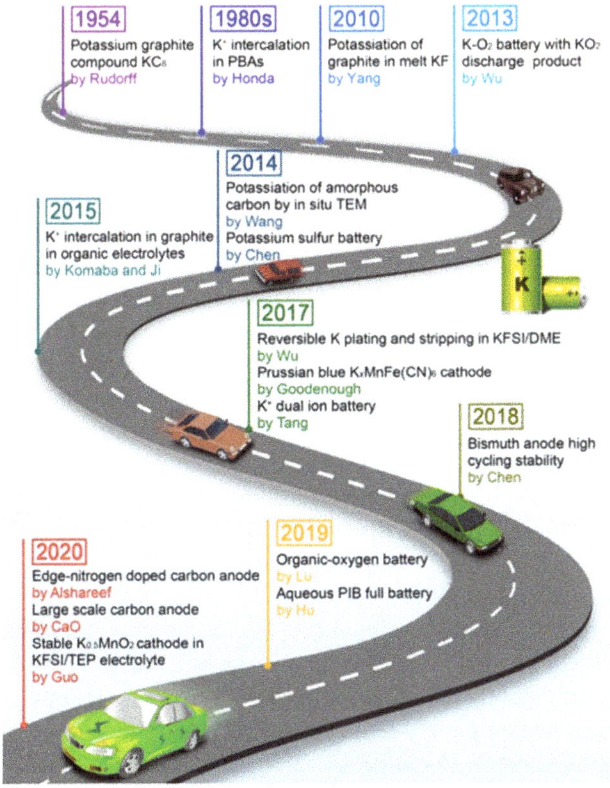

Figure 2.3. The history of key nodes in the development of rechargeable potassium-based battery. Source: [70]. Reproduced with permission of Elsevier.

people's enthusiasm for the K–S battery. Thence, the feasible integration of a K anode and S cathode in the new K–S battery is expected to generate large-scale fixed power, which can encourage energy density and cost-effective application [63, 71]. Although K–S batteries have these advantages and progress in recent years, there are also huge obstacles, so the development of K–S batteries is still in its infancy. Moreover, compared with the lithium–sulfur (Li–S) battery that belongs to the metal sulfur system, the development history of the K–S battery is shorter. In the final analysis, the performance of K–S batteries is always unsatisfactory because of the following four obstacles: (1) The notorious shuttle effect caused by soluble potassium polysulfides on the reversible oxidation/reduction reaction in the charging/discharging processes caused huge negative effects, resulting in low coulomb efficiency (CE) and severe capacity attenuation; (2) the insulating properties of a S cathode lead to its insufficient utilization; (3) K^+ ions with a large radius (1.38 vs Li^+ ions, 0.76) are difficult to be used in the graphite anode fast insertion/extraction in the medium, resulting in slow diffusion kinetics; at the same time, the difference in the density of the charge and discharge products causes the electrode volume to explode, making electrochemical reversibility worse; (4) the highly reactive potassium metal anode leads to unstable solid electrolyte interface and uncontrollable growth of dendrite [71–73].

Here, we have summarized in detail the reasonable physical design and strategies of advanced K–S batteries in recent years in terms of individual battery components, mainly including a cathode and anode. This part will start from the protection strategy of anode, including the current suppression of potassium dendrite and the attempt of liquid anode. Then, the cathode strategies including physical chemistry of carbon host, alloy addition and covalent bond of a S cathode are reviewed.

We first considered possible energy storage mechanisms to explore in depth the factors that affect the electrochemical performance of K–S batteries and promote their development. Regarding the chemical mechanism of the battery reaction, the K–S battery is similar to the Li–S battery and has undergone a series of complex conversion reactions. Although the researchers have done a lot of follow-up work for this purpose, due to the polysulfide intermediates are reconstructed during the cycle and are easy to dissolve with poor crystallinity, the basic electrochemical reaction mechanism has not been fully and clearly presented to people [74]. According to the existing literature on the discharge products of the K–S battery, the energy storage mechanism of the K–S battery could be divided into four categories, including K_2S_3 based discharge mechanism [75], K_2S based discharge mechanism [69, 76], $K_2S_2 + K_2S$-based discharge mechanism [74] and $K_2S_2 + K_2S_3$-based discharge mechanism [69].

From the perspective of thermodynamics, because the Gibbs formation energy of K_2S_3 is $\delta G_f^\circ = -528\ kJ{\cdot}mol^{-1}$ (at 25 °C), and the δG_f° of $K_2S = -410\ kJ{\cdot}mol^{-1}$, so the thermodynamics stability of K_2S_3 is higher after comparison [75]. On the other hand, K_2S_3 cannot be completely reduced to K_2S due to the slow reduction kinetics [73, 75]. Therefore, some scholars believe that K–S batteries usually work between sulfur and K_2S_3 at RT. Chen and his colleagues prepared a composite cathode composed of sulfur and CMK-3 in 2014 [75]. After charging and discharging the battery, the charging and discharging products were characterized. It is found using x-ray diffraction (XRD) that the first discharge curve is very consistent with the K_2S_3 PDF card. And the transmission electron microscopy (TEM) image displayed by the charge and discharge product also belongs to K_2S_3. What is more convincing is that the sulfur peaks in the Raman spectra of the first discharge and the first charge disappear first and then appear. According to the above XRD, TEM and Raman spectroscopy analysis, Chen proposed that the possible mechanism of the K–S battery depends on the conversion reaction: $3S + 2K \rightleftharpoons K_2S_3$. By generating K_2S_x catholyte in liquid phase, Hwang *et al* fabricated a new type of K–S battery (K||K_2S_x ($5 \leqslant x \leqslant 6$) catholyte||CNT) to explore the reaction mechanism [75]. In the initial charge and discharge process, the formation of K_2S_3 was confirmed by the results of Raman and XRD. The reversible reaction of K_2S_3 was K_2S_x ($5 \leqslant x \leqslant 6$) $\rightleftharpoons$ $K_2S_3 \rightleftharpoons K_2S_5$. Wang prepared a battery using a CMK-3/sulfur-based composite material in an electrolyte composed of potassium bistrifluoromethanesulfonimide (KTFSI)/diglyme (DEGDME) as the cathode [75]. Under the potentials of 2.3, 2.1 and 1.2 V, the sulfur-based cathode was immersed in the electrolyte for UV–visible absorption spectroscopy. Finally, the redox intermediate model was confirmed. With the help of x-ray diffraction, Wang obtained evidence for understanding the mechanism of potassium–sulfur battery chemistry. After a series of experiments, the charge and discharge mechanism is based on the reversible phase change of $S_8 \rightleftharpoons K_2S_6 \rightleftharpoons K_2S_5 \rightleftharpoons K_2S_4 \rightleftharpoons K_2S_3$. Based on the previous work in a lithium sulfur or sodium sulfur system, some researchers concluded that the discharged K_2S_3 may be reduced to K_2S_x due to disproportionation and compression reaction between polysulfides. This explanation revealed the source of high capacity of sulfur cathode. In order to deeply understand the charge–discharge mechanism of the K–S battery, UV–vis absorption spectra and x-ray photoelectron spectroscopy studies were carried out by Yu *et al* [76] After analysis, the charging/discharging process can be shown as $S_8 \rightleftharpoons K_2S_6 \rightleftharpoons K_2S_4 \rightleftharpoons K_2S_2 \rightleftharpoons K_2S$. In addition, Hwang *et al* proposed a battery composed of a sulfurized polyacrylonitrile (SPAN) composite cathode and polyacrylic acid binder, and used x-ray photoelectron spectroscopy measurement to reveal the electrochemical evolution of sulfur in the cathode and confirmed that the final discharge product is K_2S [69]. As the depth of their discharge continues to increase, the energy storage mechanism of the SPAN electrode in the K–S battery can be explained by the following equations [69].

$$-C - SK + K^+ + e^- \rightarrow -C + K_2S \tag{2.14}$$

$$-C - SK + 2K^+ + 2e^- \rightarrow -C - K^+ + K_2S \tag{2.15}$$

$$-C - K^+ + K_2S \rightarrow -C - SK + K^+ + e^- \tag{2.16}$$

In addition to the discharge products dominated by K_2S and K_2S_3 alone, there are also discharge products in which K_2S_2 is mixed with K_2S_3 and K_2S. Furthermore, the researchers also confirmed the discharge mechanism of K_2S_2 + K_2S_3 and K_2S_2 + K_2S through experiments and theoretical analysis.

In general, bare potassium metal will be used as a negative electrode of the RT K–S battery because of its low redox potential, high ion mobility and migrations number. Nevertheless, the high reactivity of potassium metal, serve electrolyte loss, the infamous growth of dendrite that is difficult to restrain, increasing the possibility of shuttle effect [77]. All of these pose great challenges to the realization of a stable potassium metal anode. Zhou *et al* limited the potassium metal to the DN-MXene/CNT scaffold to increase the stability of the potassium metal anode [78]. This porous scaffold imparts a uniformly loaded potassium metal negative electrode, and at the same time suppresses the huge volume change caused by the imbalance electroplating and peeling process of potassium. After being assembled into a K–S battery, compared with the unmodified potassium negative electrode, this negative electrode shows significantly enhanced cycling stability and outstanding bending ability. In order to suppress the high reaction activity of the K anode, Lai and his colleagues learned from the research experience of an organic oxygen battery and prepared stable biphenyl K anode for the K–S battery (BpK-S) and obtained a similar battery capacity of about 528 mAh g^{-1} [73a]. However, compared with the K–S battery, the voltage of the BpK–S battery is slightly lower. Liu and his colleagues innovatively prepared 3D copper collector with functionalized potassium surface (rGO@3D–Cu) to improve the stripping/electroplating process of potassium in KPF_6–carbonate electrolyte [68]. Specifically, graphene oxide promotes the uniform formation of metal films on 3D copper, which reduces the current density and geometrically inhibits the growth of metal dendrites. Benefitting from rGO@3D–Cu and the chemical and geometric effects of the substrate, the batteries are assembled rGO@3D–Cu at the current density of 0.1–2.0 mA cm^{-2} and has excellent cycle stability in the test of symmetrical cell. In conclusion, a potassium metal anode has received more and more attention because of its low potential (−2.93 V) and high specific capacity of ~687 mAh g^{-1}. Nowadays, researchers often use excessive amounts of potassium when assembling K–S batteries in the laboratory. One of the defects of this method is that it may lead to negative metal dendrite growing continuously, serious safety problems, huge volume change and serious parasitic reaction between the electrode and electrolyte cross section. Therefore, there is still a long way to go in the research of potassium metal anodes in the K–S system, and at the same time, a valuable field has been opened up. Researchers have made efforts in surface engineering or electrolyte optimization to solve the above problems. However, the SEI formed by the reaction of the highly active potassium metal with the electrolyte still exacerbates the shuttle problem. This problem is also the research direction for the widespread application of K metal battery that people need to pay attention to: that is to achieve a non-dendritic potassium anode, achieving a thin and tough solid electrolyte interface, protecting K metal, and allowing homogeneous K^+ plating and stripping.

The linchpin to achieving large capacity energy storage device is electrode material. Like the Li–S battery, the study of cathode materials has always been at the mainstream in the development of the K–S battery. In recent years, the research of related cathode materials is increasing, which means that the research trend is constantly clear. In this part, we are going to summarize the modification measures and electrochemical performance of positive electrode materials in the application of K–S batteries. First of all, the preparation of a carbon/sulfur composite cathode to solve the practical problems of the K–S battery has always been one of the priorities of scholars, which is mainly due to the advantages of carbon-based materials, such as rich resources, low price, stable chemical properties and so on. Moreover, carbon-based materials have good conductivity, high specific surface area and abundant chemical activity, which could provide abundant reactive sites for sulfur. At the same time, the multi-stage pore sizes are conducive to the effective adsorption of polysulfides, and the stable framework is conducive to the volume expansion of the buffer electrode. Zhao proposed the basic experiment of synthesizing CMK-3/sulfur composite in 2014 [75]. The specific process is to grind the mixture of sulfur and carbon-based materials evenly, and then keep it in argon at 160 °C for 12 h. After that, when the TEGDME–potassium perchlorate ($KClO_4$) electrolyte was used in the assembly of the battery, the specific capacity and capacity reversibility of the K–S battery were considerable. The discharge capacity of the composite was 512.7 mAh g^{-1} in the first cycle and 202.3 mAh g^{-1} in the 50th cycle. Secondly, as a new carbon-based composite electrode material-carbon/selenium sulfide (C/SeS_2) has attracted the wide attention of researchers in Li–S and K–S batteries due to its theoretical capacity of 675 mAh g^{-1} and high conductivity. Recently, Yao *et al* reported that they have revealed an advanced K–SeS_2 battery by encapsulating SeS_2 in nitrogen-doped porous carbon fibers SeS_2@NCNFs, its interconnection structure shows extraordinary advantages, such as minimizing the volume strain during the cycle, improving the rate capability, and providing an open ion/electron path [79]. Finally, the battery system shows a high reversible capacity of 703 mAh g^{-1}. Based on the research experience reported on a SPAN composite cathode in the Li–S battery, Liu and co-workers developed a SPAN/sulfur nanocomposite cathode for K–S batteries starting from the solid–phase reaction between S and PAN [76]. The specific experimental process is as follows: first heat the mixture of polyacrylonitrile and sulfur (1:4 weight) in argon to 155 °C and keep it for 1 h, and then heat it to 450 °C and keep it for 5 h. In the heat treatment process, the nitrile group in the cracked polyacrylonitrile will combine with the adjacent carbon and undergo a cyclization process. The average particle size of cathode material is about 200, which is shown by nano TEM and SEM images. XRD and Raman results show that the S element in SPAN exists in the lower order molecules except S8, and covalently combines with the carbon scaffold through S–C bond. In the electrochemical performance test, the first discharge capacity of the K–S battery was ~500 mAh g^{-1}, and the reversible capacity was 290 mAh g^{-1}, showing stable electrochemical cycling performance. Similarly, Ma and colleagues successfully fabricated a carbon capture and storage system by annealing polyacrylonitrile and sulfur (1:3, weight) in argon atmosphere at 350 °C–550 °C for 10 h [70]. When carbon chloride is utilized as positive electrode material for the K–S battery, it shows high reversible capacity of 343 mAh g^{-1} and stable cycling capacity of more than 300 cycles. Surprisingly, when a high cut-off voltage of 0.8 V is used, the cycle stability can be significantly improved.

Considering the advantages of metal sulfur battery in storage, such as the K–S battery, S has been used to combine with potassium to assemble an environmental K–S battery due to its high theoretical specific capacity (1675 mAh g^{-1}) and low-cost [80]. However, the inferior electronic conductivity of S (5×10^{-30} S m^{-1}) leads to low utilization of active materials, resulting in low capacity of the K–S battery. In addition, the shuttle effect of soluble potassium polysulfide will lead to continuous loss of S and fast degradation of properties [69]. In contrast, the electronic conductivity of selenium (Se) is much higher than that of sulfur (1×10^{-3} S m^{-1}). The theoretical specific capacity is up to 675 mAh g^{-1} and the volume capacity is 3253 mAh cm^{-3} [81]. In addition, as shown by the large amount of research on Li–S/Se batteries and Na–S/Se batteries, polysulfides produced in S-based positive electrode materials prefer to react with the solvent of electrolytes. However, the cathode based on Se usually shows one-step conversion reaction, producing fewer Se compounds, so as to prevent side-reactions and improve the stability of Se species because of their low solubility in carbonate electrolyte [82]. Therefore, Se also has more stable electrochemical properties and is highly compatible with carbonate electrolyte, so it has been used as a substitute for S, matched with metallic K for the construction of rechargeable potassium–selenium (K–Se) batteries. The first model of the K–Se battery was reported by Guo and his colleagues in 2017 [83]. Up to now, due to the short history of the development of the K–Se battery, the basic principle of the system is still not fully understood. The K–Se battery is facing a series of reasonable challenges from kinds of key issues, like the shuttle effect of selenium polyselenides (KPSs) and growth of dendrites on the surface of a K metal anode. For optimizing this new electrochemical energy storage system, people put forward some advantageous strategies, especially in the protection of an anode and cathode, so that the electrochemical performance of the K–Se battery has been significantly enhanced, and great progress is made for the K–Se battery.

The typical structure of the K–Se battery is composed of a Se cathode, separator, electrolyte and K anode. A Se cathode is mainly a complex of Se and carbon materials, yet Se exists in the complex in various molecular states, including ring-shaped Se_8, chain-like Se_n and small molecule Se, which correspond to three charge–discharge mechanisms respectively. Generally speaking, the electrochemical mechanism of the K_2Se battery is decided by the double electron reaction between Se and K: $2K + Se = K_2Se$. Therefore, the theoretical specific capacity of the K–Se battery is 675 mAh g^{-1} and the energy density is 722 Wh kg^{-1}. However, the complete reduction of a Se cathode is a complex multi electron reaction, which depends on the valent state of Se. So far, a certain number of studies have been made in determining the electrochemical mechanism of the K–Se battery, but it is still difficult for us to fully understand the electrochemistry of them.

When the selenium cathode has a ring-shaped Se_8 molecular structure, the electrochemical reaction of the K–Se battery will go through two steps in the continuous cycle process, and transform into the final product and form selenium polyselenides [84]. In this respect, the electrochemical reaction between a selenium cathode and potassium anode could be described as:

$$nSe + 2K^+ + 2e^- = K_2Se_n \tag{2.17}$$

$$K_2Se_n + 2(n - 1)K^+ + 2(n - 1)e^- = nK_2Se \tag{2.18}$$

Then, Yu *et al* recently studied the electrochemical mechanism of chain-like Se_n. It is also a two-step reduction. The results of the study on the electrochemical mechanism are written in the following equations:

$$\text{Discharge: } Se_n + 2K^+ + 2e^- = K_2Se_n; \tag{2.19}$$

$$K_2Se_n + 2(n - 1)K^+ + 2(n - 1)e^- = nK_2Se \tag{2.20}$$

$$\text{Charge: } K_2Se = K_2Se_n + 2(n - 1)K^+ + 2(n - 1)e^-; \tag{2.21}$$

$$K_2Se_n + 2(n - 1)K^+ + 2(n - 1)e^- \tag{2.22}$$

Se molecules could be split at high temperatures and further transformed into smaller Se molecules, and the formed smaller Se molecules could stably exist in the nanopores of carbon. Moreover, the small molecules of Se do not need to be gradually reduced to KPSs and finally reduced to K_2Se. They can quickly react with potassium to form K_2Se through collision kinetics. Yu and his colleagues first fabricated another short-chain small molecule (Se_2) and studied its electrochemical behavior in K^+ ion storage [85]. They performed *in situ* Raman spectroscopy to investigate the structural changes of Se_2 small molecules during the charge and discharge process. They also proved that the probability of K_2Se formation is low through DFT calculations, which means that K_2Se is thermodynamically the most stable product in K–Se batteries. Finally, an all solid state oxidation–reduction mechanism for the K–Se battery is proposed in terms of thermodynamics and kinetics. This electrochemical reactions could be described as follows:

$$\text{Discharge: } Se_2 + 2K^+ + 2e^- = K_2Se_2; \tag{2.23}$$

$$K_2Se_n + 2K^+ + 2e^- = 2K_2Se \tag{2.24}$$

$$\text{Charge: } 2K_2Se = Se_2 + 4K^+ + 4e^- \tag{2.25}$$

There are many factors that damage the life and safety of the K–Se battery, one of which is worth noting is mainly concentrated on potassium metal. First, it is still the problem of uncontrollable growth of metal dendrites. As mentioned above, metallic potassium is directly used as an anode in the rechargeable K–Se battery to ensure the supply of potassium ions. Nevertheless, due to the uneven distribution of ions/electrons in the electroplating process of potassium and selenium, dendrites may be generated, which may damage the separator and bring out internal short-circuit of the battery. K dendrites growth has become one of the crucial causes of equipment failure and safety damage of the K–Se battery. Furthermore, the possibility of potassium dendrite growth brings safety concerns, and puts forward higher requirements for scalable manufacturing and packaging technology of the battery. Potassium metal also has another drawback that can threaten the safety of the K–Se battery, which is its high chemical reaction activity which enables potassium metal to react with electrolyte and other materials in the device, and a lot of heat will be released during the reaction. At this time, if the heat and pressure continuously accumulate in the cell to the mechanical strength and softening point limit of the package, it will cause thermal runaway or package rupture. Consequently, the possibility of accidents of the whole K–Se battery will be 100%. There is no doubt that it will not only cause damage to the surrounding equipment and property, but also threaten the health and life of users or nearby people. What is more worrying is that if thermal runaway occurs in the integrated battery cluster of large-scale power storage, then the thermal runaway of one battery can quickly spread to its adjacent battery, which is likely to bring greater danger. Thus, it is necessary to strictly prove and control the use of chemical active metal potassium in order to meet the standard health and safety standards. In short, the problems of a potassium metal anode usually lead to rapid capacity decay, poor rate performance, safety problems and other serious problems, which pose a challenge to the promotion of the work of the K–Se battery. It is worth noting that these barriers are also the source of cell cycle differentiation. This shows that the Se based cathode has a gradual change in the process of potassium storage. However, the polarization phenomenon of the K–Se battery is caused by many vital factors. The first is that the dissolution of alkaline phosphatase leads to the deterioration of electrolyte environment and the occurrence of side effects of electrolyte consumption. Secondly, it is very difficult for potassium ions with large radius to migrate in the solid electrode, which increases the interfacial polarization. The solidity of SEI film will suffer from the huge volume change of a selenium cathode, which triggers the continuous formation and fall off of SEI, further aggravating the polarization of the interface. The high chemical reactivity of potassium is easy to aggravate the occurrence of parasitic reaction, which gives rise to the rapid loss of electrolyte and the sharp increase of cell polarization. For the sake of meeting the challenge of an anode side, researchers have developed several reasonable methods to eliminate the hidden danger in recent years. Liu and his colleagues reported that adding 3% FEC electrolyte additive can prevent selenium and polysselenium from dissolving from the cathode to the electrolyte by forming a stable SEI layer on the electrode surface, and protect the consistency of metal anode structure by reducing the deposition of solid selenium elements, thus benefiting the stability of the Na–Se battery [86]. Sun *et al* demonstrated the feasibility of introducing concentrated electrolyte into the K–Se battery for the first time in 2020 [80]. In the electrolyte of 5 M KTFSI/DEGDME, the key problem of low battery voltage and insufficient energy utilization are solved. As a result, the average high voltage of 1.85 V and high energy density close to the limit of 998 Wh kg^{-1} are achieved, which are superior to the current K–Se, Na–Se, and Al–Se batteries. Wu *et al* designed an artificial protective layer on the surface of a potassium metal anode by one-step method, that is, the side reaction caused by solvent and active oxygen, and SEI was composed of KF, Sb, and KSb_xF_y to prevent the corrosion of a potassium anode and maintain a great ionic conductivity [87]. Besides, the cathode is also the most challenging side of the electrochemical performance of the K–Se battery, so stabilizing the selenium-based cathode plays an important role in improving the electrochemical performance of the K–Se battery. Although selenium has greater advantages over insulating sulfur in terms of electronic conductivity, pure selenium could not be used as a positive electrode material for the K–Se battery. Therefore, it needs to be integrated into a conductive carrier material with electron transmission channels to enhance the electronic conductivity of the entire electrode [82]. Alkaline phosphatase, which is easily soluble in the electrolyte, is formed during the reversible conversion of selenium in the cathode, which leads to capacity degradation and may eventually cause the failure of the battery [88]. Thence, the prepared carrier material should have a space confinement effect, so as to effectively inhibit the shuttle effect. Generally speaking, the design of positive electrode materials should meet the following two crucial parameters: high conductivity and effective immobilization of selenium species. At present, the reasonable design of positive electrode materials for the K–Se battery mainly includes encapsulating Se within carbon materials, the synthesis and architecture regulation of Se/hollow carbon composites, combining Se and S to form Se_xS_y alloys, and the fabrication of Se/carbon composites through the molten–diffusion method [89].

2.2.3 Challenge of K anode

On the basis of potassium metal as the negative electrode, we have summarized the data published by our predecessors, and believe that the root cause of electrochemical instability is the excessive activity of potassium metal and uncontrollable dendrite growth, which causes it to be a common instability in carbonate and ether solvents. Among them, the reason for the early dendrite growth is the thermodynamic interaction between plating/stripping potassium and conventional copper current collectors. There are data in the literature that the weak bond between the potassium atom and the copper surface caused the microstructure of the potassium metal to fall off from the beginning. In terms of parasitic reactions, not only is the reaction of potassium metal with the electrolyte more than that of Li, but the reaction of Li with the current collector (according to the thermodynamics of membrane wetting) is also less. Throughout the full text, we continue to discuss key outstanding scientific issues, while pointing out related challenges and prospective directions for future research. This is done in terms of improved 'repairs' and much-needed fundamental insights, far lagging behind known Li metal anodes.

At present, potassium metal batteries are facing many difficulties (figure 2.4). For example, most of the basic mechanisms of potassium deposition kinetics have not been understood to a great extent, compared with Li plating. We are in the infancy in terms of the basic understanding of potassium's SEI structure, chemistry, formation and stability, mechanical properties and diffusion properties. Similar to the SEI based on the Li metal surface, the SEI based on the potassium metal may be a dynamic structure. In the fluctuation of the voltage, its length will grow and shrink for each electroplating/stripping cycle. However, there is no report to quantify the potassium-based SEI voltage, rate, temperature, and total capacity dependent structure. In the past few decades, researchers have launched a variety of solutions to solve the problem of lithium metal, which can also be transferred to the study of potassium metal batteries. Nevertheless, the results are not very impressive in virtue of the repeatability of the method. And the core issue is that the growth of potassium metal dendrites may be more difficult to handle than lithium dendrites or even sodium dendrites. Furthermore, there has been a lack of advanced *in situ* specific analysis tools used to further understand the local structure of potassium-plated metals and related SEIs. Yet, it should be noted that compared with Li, potassium is more reactive with air,

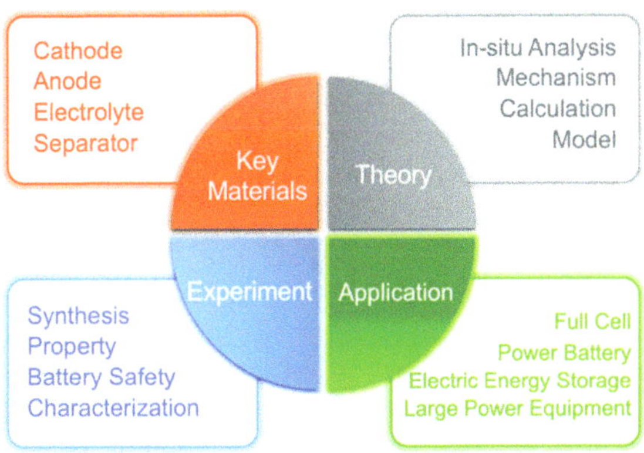

Figure 2.4. Potassium-based batteries are facing dilemmas that need to be resolved. Source: [92]. Reproduced with permission of The Royal Society of Chemistry.

water vapor and most electrolytes. Potassium also has a lower melting point than lithium, making it more susceptible to thermal induced artifacts during sample preparation. But potassium is heavier, making it less susceptible to thermal electron beam damage, such as damage caused by impact and sputtering. If these work are really carried out, it may reveal unexpected differences or similarities in terms of overall SEI formation/metal plating phenomenology.

2.3 Mg anode for Mg-ion batteries

2.3.1 Magnesium properties

Due to high energy density and long cycle life, lithium-ion batteries have greatly promoted the development of portable devices and electric vehicles and dominated the field of energy storage in recent years [90]. However, the reserves and distribution of lithium and cobalt resources, as well as the safety issues of the organic electrolyte, have limited the large-scale application of lithium-ion batteries [91]. At the same time, the growing demand for sustainable energy and reducing carbon emissions is driving the transfer of battery technology to earth-rich materials and the development of low-cost batteries. With these key issues in mind, there is growing concern about sustainable electrode materials and lithium-free technologies.

Magnesium is one of the 10 commonly used non-ferrous metals that is abundant in the Earth's crust, with a content of 2.1%–2.7%, ranking sixth among all elements, and fourth only to aluminum, iron and calcium [92]. Magnesium, first identified by Joseph Black in 1755, is located in Group IIA of the periodic table. In 1808, David of England reduced magnesium oxide with potassium to produce magnesium metal, which has a silvery white light alkali earth metal, chemically active and can react with acids to produce hydrogen, has a certain ductility and thermal dissipation. Magnesium is widely distributed in nature and is one of the essential elements in the human body.

Using magnesium as an anode has the following advantages:

1. High safety. Magnesium is more chemically stable than lithium. Lithium has a melting point of about 180 degrees, while magnesium has a melting point of 650 degrees, which makes it safer than lithium [93]. In addition, compared with lithium, the magnesium anode is safer due to its non-dendritic deposition and low air sensitivity [94].
2. Rich reserves of raw materials and low production cost. Magnesium is also much more abundant than lithium (2.1%–2.7% of the Earth's crust, while lithium is $7\times10^{-4}\%$) [95]. It can be readily available in a variety of minerals, for example, dolomite ($CaCO_3 \cdot MgCO_3$) or magnesite ($MgCO_3$). Seawater also contains a lot of magnesium salts. So it's cheaper to use magnesium as anode electrode [96].
3. High theoretical specific capacity. Magnesium, the second major group of the third period, has an electron valence of +2, while lithium has an electron valence of +1. Theoretically, magnesium has twice as much electronic valence as lithium [96]. As shown in (figure 2.5) Mg also has a low reduction potential (−2.356 vs SHE) and a high theoretical mass

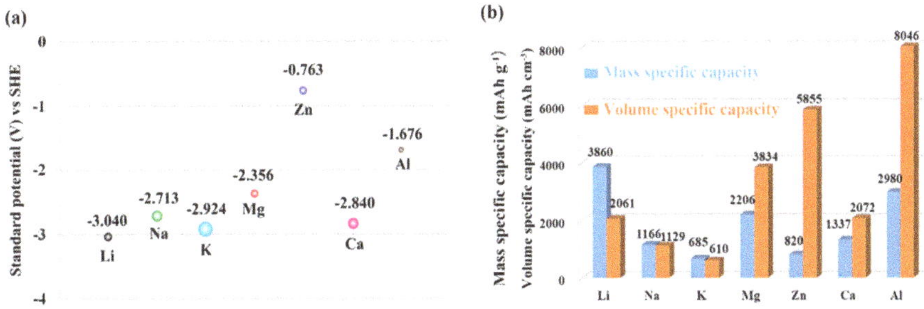

Figure 2.5. (a) A comparison of monovalent/multivalent metals in the standard potential and ionic radius; (b) A comparison of monovalent/multivalent metal-ion batteries in the theoretical capacity.

specific capacity of 2206 mAh/g and a high theoretical volume specific capacity of 3834 mAh cm^{-3}. So, a magnesium anode has high theoretical specific capacity that can store more energy than lithium-ion batteries, allowing products such as smartphones and computers to have longer battery life.

Based on the above advantages, a magnesium based chemical power source has been widely considered and is a strong competitor of the next generation of green power source.

2.3.2 Development of magnesium anode

The types of magnesium batteries mainly include magnesium dry battery, seawater activated magnesium battery, magnesium–air battery, magnesium solid state battery and magnesium secondary battery [97]. These types of batteries are briefly described below.

2.3.2.1 Magnesium dry battery

The magnesium dry battery was developed in 1928 by Wood Robert of the American Magnesium Industry Corporation and were first used as Army communications batteries in 1968 [98]. The magnesium dry battery mainly uses magnesium and its alloy as the anode, MnO_2 as the cathode, and $Mg(ClO_4)$ as the electrolyte ($Mg|Mg(ClO_4)_2|MnO_2$) [98].

Reactions involved in batteries

$$\text{Anode: } 2Mg + 2OH^- \rightarrow Mg(OH)_2 + 2e^- \tag{2.26}$$

$$\text{Cathode: } 2MnO_2 + H_2O + 2e^- \rightarrow Mn_2O_3 + 2OH^- \tag{2.27}$$

$$\text{Overall: } Mg + MnO_2 + H_2O \rightarrow Mg(OH)_2 + Mn_2O_3 \tag{2.28}$$

Compared with the traditional (Zn/MnO_2) dry battery, the Mg/MnO_2 dry battery has several outstanding advantages, that is, long life, good capacity retention, wide operating temperature and high thermal stability. Several bottlenecks restricting the development of MG primary batteries mainly include the following aspects:

(1) The low corrosion resistance of magnesium anode material to various common electrolyte components leads to low utilization rate of anode electrode and causes low actual energy density, which hinders its wide application.
(2) Due to the active chemical properties, magnesium metal is easy to form a passivation film in the air or humid conditions, which is mainly composed of some insoluble magnesium oxides such as magnesium oxide, magnesium hydroxide and magnesium carbonate and other magnesium salts. The growth of the passivated surface film blocks the transmission of ions and electrons, resulting in serious voltage hysteresis phenomenon.

These issue involved in the application bottlenecks can be divided into three categories: the low kinetics of active materials, the passivation of $Mg(OH)_2$ film formed on the surface of the Mg anode electrode, and the high corrosion rate of the anode electrode caused by inappropriate electrolyte. Therefore, the overall performance of Mg/MnO_2 cells can be greatly improved by increasing the electrochemical activity of the electrode materials (i.e., Mg and MnO_2) and using appropriate electrolyte.

2.3.2.2 Seawater activated magnesium battery

Seawater activated batteries were first introduced in the 1940s to meet the need for high-energy density, long-life batteries with excellent low-temperature performance for military applications. The batteries can be designed for a wide operating cycle ranging from a few minutes to a few days, as well as for low current and high current, pulse and constant operation batteries [99]. Because seawater is used as a non-pressurized storage condition for electrolytes and electrodes, and the power density can be very high, these advantages offer higher practicability and lower cost for seawater activated batteries compared to lithium or alkaline batteries used in subsea equipment. Therefore, this battery can be used for a variety of battery underwater applications, such as life buoy, sonobuoy, torpedo, life raft and detection device.

Usually, the seawater activated battery consists of three parts, the active metal as the anode, metal halides such as AgCl, CuCl or CuI as the cathode, and seawater as the electrolyte. The cells rely on the dissolution of the active metal anode in seawater and the reduction of the metal halide in the positive electrode to produce an electrochemical potential. Therefore, the key material used in seawater batteries is the metal active metal electrode, which increases the battery voltage and minimizes self-discharge. Magnesium is the main anode material because of its anode electrode potential, strong discharge activity and high discharge potential. Several seawater activated magnesium alloy negates have been developed for military and commercial applications over the past few decades, but many designs are apparently unpublished or patented.

The first commercially developed seawater activated battery was the magnesium/silver chloride (Mg/AgCl) system developed by Bell Laboratory [99]. Due to its excellent high rate discharge performance and long shelf life, it was applied to battery-powered torpedoes. The main chemical reactions of the battery are as follows:

$$\text{Anode: Mg} \rightarrow \text{Mg}^{2+} + 2\text{e}^- \quad (2.29)$$

$$\text{Cathode: 2AgCl} + 2\text{e}^- \rightarrow 2\text{Ag} + 2\text{Cl}^- \quad (2.30)$$

$$\text{Overall: Mg} + 2\text{AgCl} \rightarrow \text{MgCl}_2 + 2\text{Ag} \quad (2.31)$$

Since then, seawater activated magnesium cells based on this technology have been widely used in small equipment, such as acoustic waves, radio buoys, lifesaving equipment, ocean markers and emergency lights, and meteorological radiosondes, due to their remarkable performance in all aspects. Due to the high cost of silver and the impracticality of recovering it after use, other non-silver batteries have been developed. At present, the systems that have been successfully developed and used are $Mg/CuCl_2$, $Mg/PbCl_2$, Mg/Cu_2I_2-S, Mg/CuSCN-S and so on. However, any of these systems can hardly compete with all the characteristics of the Mg/AgCl system.

There is also a side reaction between the magnesium anode and the aqueous electrolyte, resulting in the production of magnesium hydroxide, hydrogen gas and release heat.

$$\text{Mg} + 2\text{H}_2\text{O} \rightarrow \text{Mg(OH)}_2 + \text{H}_2 \quad (2.32)$$

The generation of gas-producing and non-conductive by-products affects the stability of the battery. In the future, the development of corrosion resistant magnesium anode will be an important direction to improve the performance of seawater activated batteries.

2.3.2.3 Magnesium–air battery

Metal–air batteries have attracted much attention due to their high theoretical energy density and low cost, and have been considered as a promising electrochemical energy storage and conversion device [100]. Among various types of metal–air batteries (mainly including lithium–air and zinc–air batteries), Mg–air batteries use neutral saline electrolytes instead of caustic solutions, which has great significance for civil use, and has received more and more attention. It is composed of a Mg (or Mg alloy) anode, air cathode and salt electrolyte.

The theoretical voltage of a Mg–air cell is 3.1 V and the specific energy density reached 6.8 $kWh{\cdot}kg^{-1}$.

In general, current Mg–air batteries are non-reusable batteries. In fact, the Mg–air battery can be 'refuelable' by replacing the used Mg anode and electrolyte with a new Mg anode and electrolyte, so that the Mg–air battery can be mechanically reused. This has great significance to solve the problem of 'refueling' of the electric car and worth further study by researchers.

The issues of Mg–air batteries are high polarization and low coulomb efficiency. For example, the operating voltage is usually less than 1.2 V, and the actual specific energy is less than 10% of the theoretical value. There are two reasons for this problem: the corrosion of Mg anode electrode due to the reaction between Mg and electrolyte, and the kinetics of slow oxygen reduction in the cathode electrode of air.

2.3.2.4 Rechargeable Mg-ion batteries

A magnesium secondary battery is a new type of rechargeable battery developed in recent years. Rechargeable Mg-ion batteries (RMIBs) have many advantages such as low cost, safety and environmental friendliness [101]. Its structure is shown in figure 2.6, the core of RMIBs consists of Mg anode, electrolyte solution and cathode that Mg^{2+} can insert or extract. The working principle of RMIBs is similar to the lithium-ion batteries. During the process of charging or discharging, Mg^{2+} is constantly de-inserted or extracted from the cathode accompanied by the deposition and dissolution of Mg^{2+} in the anode.

As early as 1917, Nelson *et al* found that magnesium could have reversible dissolution and deposition in the ether solution of Grignard's reagent. Genders *et al* used microelectrodes to study the basic electrochemical reactions of magnesium in tetrahydrofuran (THF) and propylene carbonate (PC), and found that magnesium could be deposited or dissolved on copper with high coulomb efficiency in the THF solution dissolved with Green's reagent [103]. However, due to the narrow potential window of Grignard's reagent, it cannot be used in the magnesium secondary battery system.

In 1990, a complete magnesium secondary battery system was reported for the first time by Gregory *et al*, which used $Mg(Br_4)_2$ (R as an organic group) solution with Mg_xCoO_y as the cathode [104]. The results show that these solutions have a wider electrochemical stability window than Green's solution, and the coulomb efficiency of charge and discharge can reach 99%. However, due to the lack of oxidation stability of the electrolyte, the utility of the battery has not been proved.

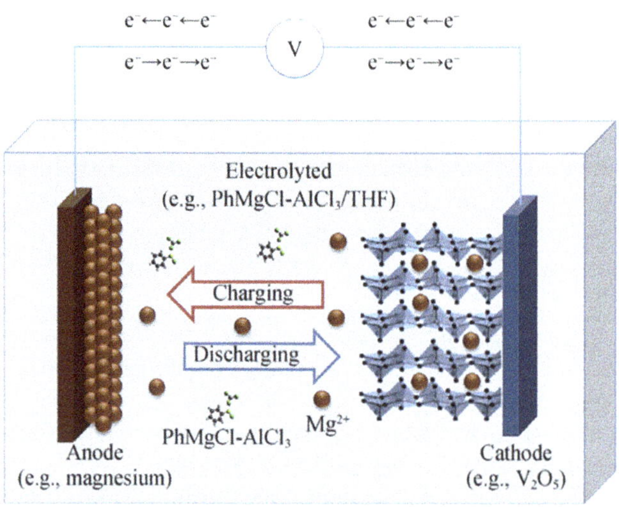

Figure 2.6. Schematic diagram of typical electrochemical reaction process of MIBs with V_2O_5 as cathode [102]. Source: [102]. Reproduced with permission of Springer.

Until in 2000, Aurbach *et al* developed a new magnesium secondary battery system, which made a breakthrough in the study of the magnesium secondary battery [105]. The system uses magnesium anode, $MgMo_3S_4$ as cathode and THF solution of $Mg(AlCl_2BuET)_2$ as electrolyte. At room temperature, the deposition/dissolution efficiency was 100%, and the theoretical capacitance was 135 Wh kg^{-1}.

Subsequently, Aurbach *et al* made further improvements to the magnesium secondary battery based on the original battery system [105]. The problems with previous battery systems were that the millimeter-size Chevrel compound $Mg_{0-2}Mo_6S_8$ positive electrode was limited by kinetics and the electrolyte had a narrow electrochemical window. The new system in the cathode material, based on the original composition of the element Se, was nanometer $Mo_6S_{8-Y}Se_y(0<y<2)$ as the inserted cathode. $AlCl_3$ and PhMgCl (1:2) were used as electrolyte. As a result, the system has a wider electrochemical window, better electrical conductivity and faster magnesium-ion insertion/deinsertion rates than the original system.

Polymer electrolyte has been used in lithium batteries due to its advantages of easy preparation, strong plasticity and high safety. There are also related studies on magnesium batteries. In 1999, Girish Kumar *et al* first used polymer electrolytes in magnesium secondary batteries and verified the reversibility of the Mg/Mg^{2+} in the electrolyte [106]. Japan Nobuko *et al* synthesized (PEO-PMA)/PEGDE/Mg $[(C_3FSO_2)_2N]_2$ polymer electrolyte with the best conductivity greater than 10^{-4} S cm^{-1} [107]. Mg was used as the anode and V_2O_5 as the cathode with 100 mAh g^{-1} specific capacity.

In 2003, the Aurbach group reported a solid-state magnesium secondary battery and prepared PVDF/tetraethanol-dimethyl ether/$Mg(AlCl_2BuET)_2$ solid electrolyte [108]. The conductivity of the electrolyte reached 3.7 S cm^{-1} at 25 °C. When using AZ31 alloy as anode, Mo_6S_8 as cathode, the battery has a discharge capacity of 115 $mAhg^{-1}$ at temperatures above 60 °C.

At present, the research of magnesium secondary battery is still in the preliminary stage, and it is still a long way from practical application. Due to the large charge density of the magnesium, magnesium secondary battery research is restricted by the following aspects of:

(1) The chemical activity of magnesium is higher, surface passivation films are readily formed in most solvents, which hinder Mg^{2+} move.
(2) Mg^{2+} have high charge density and serious solvation, so it is difficult to insert into the general matrix, which limits the choice of cathode materials.

2.3.3 Challenge of magnesium batteries

Compared with the primary battery, the secondary battery is recyclable, energy saving and has environmental protection, which is more in line with the future development direction. Magnesium and lithium lie diagonally on the periodic table, so they share many similar chemical properties. With low price, no pollution to the environment and large specific capacity, magnesium batteries have been gradually favored by people as a new battery system.

Magnesium metal has been deemed as the ideal negative material for magnesium batteries due to its low-cost, it is environmentally benign and has a low oxidation potential. But the active magnesium metal is easy to react with the electrolyte during the battery operation and generate a passivation layer. According to literature reports, there are three main methods at present proposed to overcome the remaining challenges.

2.3.3.1 Nanostructured magnesium anode

Reducing the particle size of magnesium metal could effectively improve the contact area between the electrolyte and magnesium, and then effectively improve the utilization rate of magnesium [109]. Chen and colleagues synthesized magnesium nanospheres with different particle sizes by adjusting the temperature [109]. As shown in figure 2.7(a), as the deposition temperature increases from 240 °C to 270 °C, the particle size of the nanospheres increases from 1.0–5.0 to 3.0–6.0 μm. When the temperature increased to 300 °C, the particle size of the material decreased, and the specific surface area of the material reached 1.92 m^2 g^{-1}. At this temperature, the material exhibits excellent electrochemical activity (figure 2.7(b)) and highest specific capacity (figure 2.7(c), 768 mAh g^{-1}).

2.3.3.2 Magnesium alloy anode

Compared with Mg, Mg alloys in which Mg is formed with IIIA, IVA and VA group elements show excellent compatibility with the electrolyte. Therefore, Mg alloys are also a promising research direction.

Si, Ge and Sn are common alloying elements. For instance, Nguyen *et al* using mechanical alloying process synthesized bulk Mg_2Sn. Scanning electron microscropy (SEM) of the synthesized Mg_2Sn are shown in figure 2.8(a). The particles sizes are about 720 nm. The presence of O atoms is associated with surface oxidation

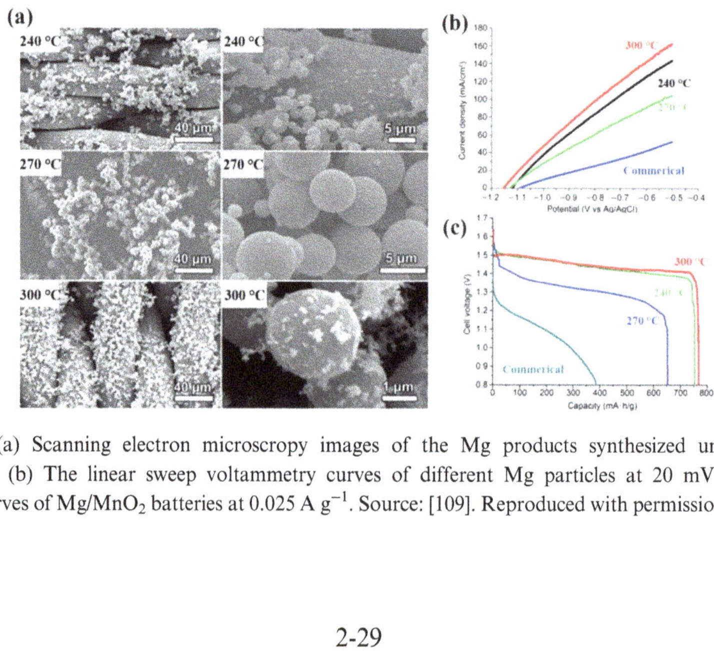

Figure 2.7. (a) Scanning electron microscropy images of the Mg products synthesized under different temperature. (b) The linear sweep voltammetry curves of different Mg particles at 20 mV s^{-1}. (c) The discharge curves of Mg/MnO_2 batteries at 0.025 A g^{-1}. Source: [109]. Reproduced with permission of Springer-Verlag.

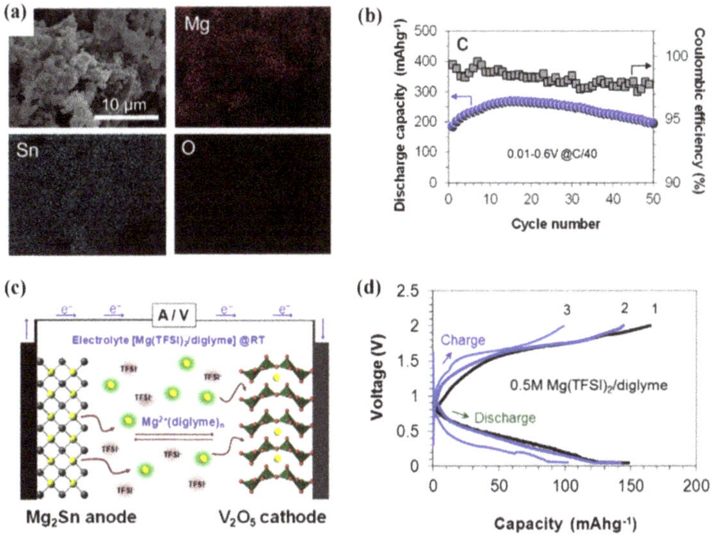

Figure 2.8. (a) Elemental mapping results of Mg_2Sn.(b) Cycling performance of Mg_2Sn anode. (c) Schematic of $Mg_2Sn//V_2O_5$ full-cell. (d) Voltage profiles of $Mg_2Sn//V_2O_5$ battery. Source: [110]. Reproduced with permission of Elsevier B.V.

during the ball milling and drying process. A $Mg//Mg_2Sn$ half-battery has the first discharge capacity of 182 mAh g^{-1} at room temperature (figure 2.8b). The structure of $Mg_2Sn//V_2O_5$ full battery is shown in figure 2.8(C). The initial coulomb efficiency is 90%, which shows excellent reversibility (figure 2.8(d)). The reaction mechanism of the battery is shown in the formula equation (2.36) and (2.37).

$$\text{Anode: } Mg_2Sn \leftrightarrow Mg_{1.5}Sn + 0.5\,Mg^{2+} + e^- \tag{2.36}$$

$$\text{Cathode: } V_2O_5 + 0.5\,Mg^{2+} + e^- \leftrightarrow Mg_{0.5}V_2O_5 \tag{2.37}$$

2.3.3.3 Artificial SEI

Different from the solid electrolyte interface (SEI) transporting Li^+ in Li batteries, a passivation layer hindering Mg^{2+} conduct would be formed induced by the decomposition of electrolyte components or parasitic reactions between electrolytes and anodes [111]. $Mg(TFSI)_2$, one of the very few Mg salts possessing reasonable solubility in ether solvents, is highly resistant to oxidation and provides a wide operating voltage range. Unfortunately, $Mg(TFSI)_2$-based electrolytes usually show a poor electrochemical performance as a result of the decomposition of Mg salt forming a MgF_2-rich blocking layer, which impedes reversible Mg plating/stripping (figure 2.9(a)) [111]. Therefore, creating an artificial SEI on the surface of magnesium that can allow Mg^{2+} to pass seems to be a very effective method. However, it should be noted that high current density and accidental damage often lead to a loose structure of the artificial film [112]. As a result, the protection layer may crack or even partially peel off so that the bare region would be

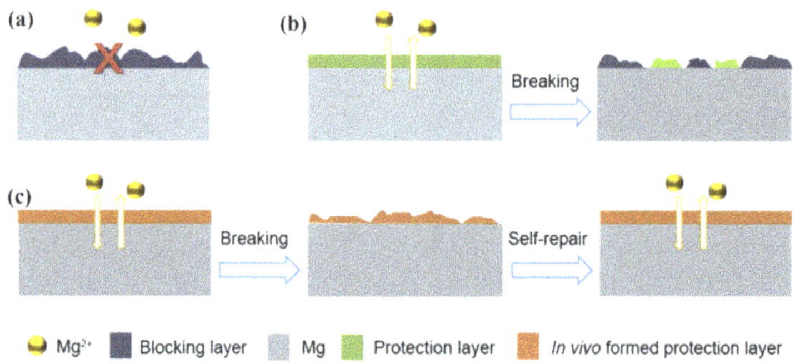

Figure 2.9. Schematic of the (a) passivation layer in blank electrolyte, (b) breaking process for protection layer and (c) breaking and self-repair process for *in vivo*-formed protection layer during Mg plating/stripping in conventional organic electrolyte. Source: [111]. Reproduced with permission of Elsevier B.V.

passivated again due to the direct contact between the Mg anode and electrolyte (figure 2.9(b)) [111].

Therefore, the ideal artificial SEI can not only allow Mg2+to pass smoothly, but also repair the damaged structure under high current density and accidental damage. For example, Zhang and co-workers propose a facile surface chemistry approach to accomplish a Ge-based protection layer *in vivo* on the surface of Mg metal by adding $GeCl_4$ into $Mg(TFSI)_2$/DME electrolyte [111]. The Ge-based protection layer provides a pathway for rapid Mg^{2+} transport and meanwhile prevents the formation of the passivation film on the Mg anode surface. On account of excess additives in electrolyte, a self-repair process would occur when the artificial protection film ruptures abruptly (figure 2.9(c)). After 3 mAh cm^{-2} Mg plating using 0.4 M $GeCl_4$ in 0.5 M $Mg(TFSI)_2$/DME electrolyte, the existence of protective layer can be seen from the SEM diagram (figure 2.10(a), 2.10(b)). As displayed in figure 2.10(c), owing to the Ge-based protection layer formed *in vivo*, the cell delivers significantly enhanced reversibility and cycling stability in modified electrolyte, with a reversible capacity of 87.8 mAh g^{-1} over 30 cycles at a rate of 10 mA g^{-1}. Predictably, with self-standing Ti_3C_2@CTAC cathode, the cells in modified electrolyte deliver excellent stability with a high discharge capacity of ~100 mAh g^{-1} even at 50 mA g^{-1} (figure 2.10(d)). This strategy paves a new way towards rational surface modification of monovalent or multivalent metal anode in simple-salt-based organic electrolytes for rechargeable metal batteries.

Although there are three ways to improve the performance of magnesium anode, it should be noted that these studies are still in their early stages and there are still a number of problems. For example, reducing the size of the active substance increases the availability of magnesium, but reduces the specific volume capacity of the metal electrode [113]. Although the alloyed electrode has a lower requirement for electrolyte, the large volume change of alloyed electrode in the cycle process will reduce the cycling stability of the electrode, while the artificial SEI membrane has poor stability and other defects under the large current density and voltage [113, 114]. These problems restrict the development of magnesium anode and need further research.

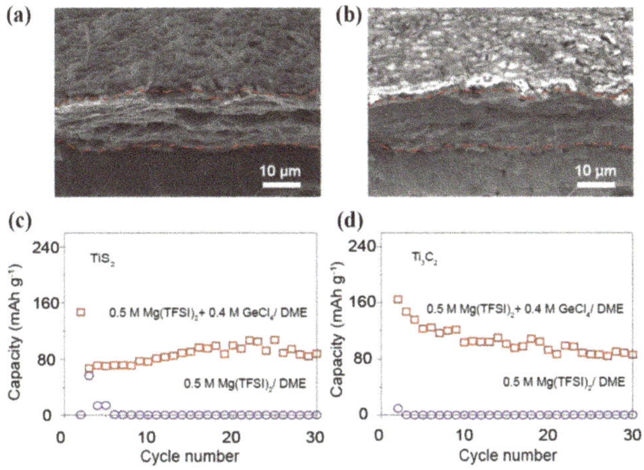

Figure 2.10. Full cell performance. Cross–section SEM images of protected anode after 3 mAh cm^{-2} Mg plating using 0.4 M $GeCl_4$ in 0.5 M $Mg(TFSI)_2$/DME electrolyte in symmetric Mg cell in (a) secondary electron mode and (b) backscattered mode. Cycling performance of (c) Mg/TiS_2 and (d) Mg/Ti_3C_2 full cell. Source: [111]. Reproduced with permission of Elsevier B.V.

2.4 Al anode for Al batteries

2.4.1 Al properties

Nowadays, the demand for energy storage is obviously rising in our daily lives. Meanwhile, in order to achieve the basic balance between peak energy production and consumption, the energy production process is gradually replacing the traditional fossil fuel by renewable energy, such as solar and wind energy. However, the intermittent nature of solar and wind energies greatly hinder their further development. Fortunately, electrochemical energy storage (EES) devices are promising feasible and high-efficiency solutions to store electric energy obtained from renewable energies. Among various EES devices, lithium-ion batteries (LIBs) have been intensively investigated and used in various applications such as portable electronics, electrical vehicles and smart grids, due to their low self-discharge rate, long cycle life and they are environmentally benign [115]. However, the massive consumption of expensive and rare minerals containing lithium and cobalt has led to restrictions on the large-scale application of LIBs [116].

In this context, many other metal-ion batteries are being explored nowadays, and the scientific interest in alternative more abundant metals (figure 2.11(a)) such as Na, Zn, K, Ca, Mg, and Al have increased in the last few years. Among them, the Al battery is considered to be one of the next generations of secondary batteries with great potential and prospect. First of all, the gravimetric capacity of an aluminum anode is comparable to Li, and its volumetric capacity is four times Li. Second, Al is abundant in the Earth's crust. Third, Al is much safer than Li as an anode. Fourth, Al^{3+} is a trivalent charge carrier that can store three times as much charge per ion transfer as Li^+. However, as a metal anode, there are many problems with Al to be solved. For example, the complexity of trivalent Al^{3+} embedded in the cathode of

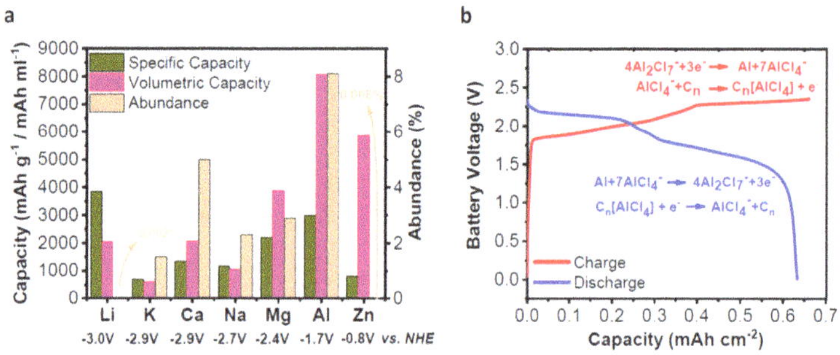

Figure 2.11. (a) Comparison of specific and volumetric capacity, redox potential and abundance of metals, which have been proposed to be promising negative electrode materials for batteries. (b) Typical charge and discharge curves of an Al-ion battery using graphitic material as cathode materials. Source: [121]. Reproduced with permission of Royal Society of Chemistry.

AIBs and the corrosion of Al anode in aqueous electrolyte hinder the application and development of AIBs to some extent [117].

2.4.2 Development of Al anode

Aluminum-ion batteries (AIBs) are usually composed of an aluminum metal foil as anode and negative current collector, chloroaluminate ionic liquid electrolyte, a graphite material as cathode and a positive current collector [118]. In the charging process (figure 2.11(b)), Al contained in the electrolyte in the form of $Al_2Cl_7^-$ is reduced to the surface of the anode to form an electrodeposited Al metal film and $AlCl_4^-$ anions in the electrolyte are released after reaction (2.38). At the same time, $AlCl_4^-$ anions contained in the electrolyte are intercalated into the crystal structure of graphite at the cathode, following the reaction (2.39). The reactions are reversed in the discharge process.

$$\text{Anode: Al} + 7\text{AlCl}_4^- \xLeftrightarrow{\text{ChargeDischarge}} 4\text{Al}_2\text{Cl}_7^- + 3e^- \tag{2.38}$$

$$\text{Cathode: C}_n[\text{AlCl}_4] + e^- \xLeftrightarrow{\text{ChargeDischarge}} \text{AlCl}_4^- + \text{C}_n \tag{2.39}$$

Unlike other battery types, such as LIBs, in which the ionic concentration in the electrolyte could keep constant in any charging state, in AIBs the ionic concentration will change during the charge and discharge process. Hence, it has been suggested that this technology should be called $AlCl_3$-graphite batteries [118], instead of AIBs, which they have been commonly called until now [119].

2.4.3 Challenge of Al anode

The total specific capacity of the active material relies on the specific capacity of the cathode and anode (equation (2.40)):

$$C_T = \frac{C_A C_C}{C_A + C_C} \quad (2.40)$$

In this formula, C_T, C_A and C_C are the total specific capacity of the battery, the anode and the cathode, respectively, and the unit is mA h g^{-1}. Unlike a rocking chair battery, it is important to note that the C_A in AIBs does not come from the Al anode but from the Al which is initially dissolved in the electrolyte. Hence, the capacity of the anode is associated with the volume and molar ratio of the electrolyte [120].

With over 20 000 cycles, the long cycle life of AIBs has been achieved. Although some breakthroughs have been made, there are still many factors to be considered for future applications in real life, such as lower rates (0.5–1 C), bigger areal capacity (> 1mAh cm^{-2}), longer cycling period (> 1 year), room temperature and in real cell conditions. At the same time, due to the lack of systematic and in-depth study on the irreversible side effects of AIBs, researchers have no way to further optimize its cycle performance. For example, the decomposition of electrolytes often has a detrimental effect on their cycling performance for LIBs, so researchers can use the coulombic efficiency to quantify the effect of side reactions of electrolyte on cycling performance (figure 2.12(a)). For AIBs, the researchers find that the coulomb efficiency of the electroplating/stripping process is less than 100% when using an aluminum-free collector (figure 2.12(b)), demonstrating that reduction occurs in the electrolyte at the anode. But researchers do not yet understand the mechanism of degradation and reduction products. Though products from side reactions at the anode may be less corrosive, cross-talking and gas generation between electrodes are still significant parameters for AIBs' performance in the sealed configuration using minimum amounts of electrolyte. All in all, the side reactions occurring at the interface of both anode and cathode need to be investigated more intensively to guarantee and improve the cycling ability of AIBs in the industry-related conditions [121].

In general, the specific power calculations for AIBs reported in the literature take into account the fact that the cathode alone leads to significantly higher values. When the mass of both positive electrode and negative electrode are considered, the specific power is decreased to about 4 kW kg^{-1} [122]. Meanwhile, the power density

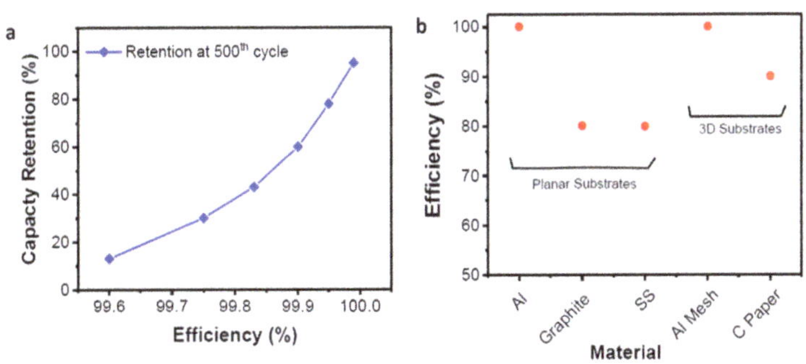

Figure 2.12. (a) Expected capacity retention of a Li-ion battery after 500 cycles as a function of the coulombic efficiency. (b) Coulombic efficiency of the plating/striping process of Al on different substrates. Source: [121]. Reproduced with permission of Royal Society of Chemistry.

of lead-acid and LIBs are ~0.3 kW kg^{-1} and ~1 kW kg^{-1}, respectively. As a result, the power density of AIBs is obviously higher than other existing mature battery technologies, which could make AIBs more appropriate to high-power applications. In addition to cycle life, aluminizing/stripping kinetics as well as mass transfer of anions through porous cathode are also critical for future AIBs research.

AIBs have higher safety than other battery technologies, for example, LIBs. Aluminum does not react fiercely with air. Furthermore, the $AlCl_3$:EMImCl electrolyte is a highly thermal stable as well as nonvolatile ionic liquid that can allow the battery to operate over a wide range of temperatures, ranging from −20 °C to 80 °C [123] However, in addition to the potential safety hazard caused by the reaction between air and metal, the chlorine produced during charging of the positive electrode will cause some potential safety concern as it accumulates continuously. Therefore, in-depth and systematic understanding of the electrode interface side effects are crucially important.

Figure 2.13(a) shows the cost of different components per kWh of AIBs, which represents a significant upfront investment per kWh (1400 USD kWh^{-1}) compared to other mature battery technologies (250 USD kWh^{-1} for LIBs). In addition to the effect of the cost of batteries, electrolyte and collector are two important factors for AIBs. Electrolytes are usually formed by mixing $AlCl_3$ and EMImCl. In general, the cost of the electrolyte is calculated to be 35 USD L^{-1}, which is almost twice the price of LIBs electrolyte (17 USD L^{-1}). Also, because AIBs have a lower energy density than LIBs, they require more electrolyte per kWh. Surprisingly, the most advanced AIBs' main contribution to the cost of the battery is the cathode collector. Ni, Cu, steel and other materials can be applied in conventional collectors for other batteries, but the anodic dissolution will occur in the electrolyte, which cannot work properly in AIBs. Therefore, more suitable and inexpensive metal anode materials are the basis of large-scale application of AIBs.

An important aspect of battery cost that is often overlooked in the AIB field is the specific areal capacity (mAh cm^{-2}). This parameter also has a significant impact on the cost of the battery, as it determines the total area of the battery and directly

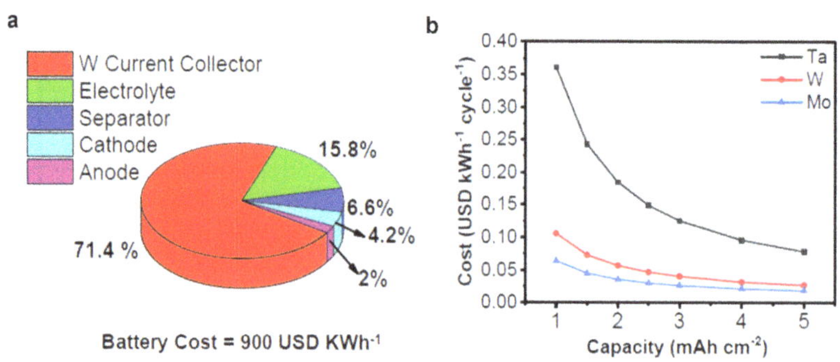

Figure 2.13. (a) Contribution of the different elements of an AIB to the battery cost for an areal capacity of 1 mAh cm^{-2}. (b) Energy cost per cycle for Al-ion Batteries vs areal capacity using different current collectors. Source: [121]. Reproduced with permission of Royal Society of Chemistry.

affects the amount of inactive material, such as the collector. Figure 2.13(b) shows how the cost per kwh and per cycle varies with the specific areal capacity. The increase in areal capacity reduces the amount of inactive material and the battery specific area, thus reducing the cost. At present, the most areal capacity reported are usually below 1 mA h cm^{-2}. Therefore, in order to make AIBs more cost competitive, researchers should also make more effort in the AIBs with higher areal capacity [122].

2.5 Zn anode for Zn batteries

2.5.1 Zn properties

Some renewable energies such as solar and wind energy have been vigorously developed given that the climatic deterioration and environmental pollution have become increasingly serious. However, these renewable energy resources are difficult to be directly used due to their discontinuous, unpredictable and unstable electricity generation. Thus, it has raised urgent demand for the development of safe, economic, efficient and sustainable EES systems [124]. Among them, the aqueous zinc-ion battery (AZIB) has been widely reported, and Zn metal is the most common anode owing to its intrinsic advantages, including high theoretical capacity (820 mAh g^{-1}), suitable electrochemical potential, and abundant natural resources [125]. In addition, zinc metal has some other outstanding advantages which make ZIBs superior to LIBs: (1) the amount of zinc on Earth is three times that of lithium; (2) ZIB assembly does not need an inert environment, which greatly reduces the cost of battery assembly; (3) zinc metal can be employed directly as anode material above other metals (Li, Mg, Ca, Al), because zinc metal shows desirable stability and reversible electrodeposition properties in the aqueous electrolyte; (4) Zn metal as anode presents high theoretical volumetric (5855 mAh cm^{-3}) and gravimetric capacity (820 mAh g^{-1}) (5) zinc and its salts are environmentally friendless, which ensures the high security of ZIBs. In recent years, zinc has been employed in various batteries, such as Zn–MnO_2, Zn–NiOOH, Zn–Ag_2O, Zn–air batteries, etc [126]. It should be noted that the reaction mechanisms in ZIBs are unclear and controversial. Three mechanisms have been demonstrated for the manganese dioxide cathode: (1) Zn^{2+} insertion/extraction reactions; (2) H^+/Zn^{2+} co-insertion/extraction reactions; and (3) conversion reactions.

2.5.2 Development of Zn anode

2.5.2.1 Nickel–zinc battery

Nickel and zinc are relatively easy to find metals. Compared to NiMH batteries, which require a lot of rare earth elements, Ni–Zn batteries require mainly zinc in addition to nickel, which is easier to find than rare earth elements. Indeed, more than a century ago, Tomas Edison recognized the potential of nickel–zinc batteries and studied them. Nickel–zinc batteries did not develop well because of the technology available at the time. In early 2010 PowerGenix, a little-known US company, announced the launch of a new nickel–zinc rechargeable battery. PowerGenix took this ancient technology and retrofitted it, especially with improved electrolyte and electrode formulations, to produce the new generation of nickel–zinc rechargeable batteries we see today.

The main preparation methods of zinc anode are paste method, electrodeposition method, compression method, chemical method and sintering method. Zinc anode can be made into charge state and discharge state, mainly depending on if the initial material is ZnO (charge state) or Zn (discharge state). Generally, the zinc anode is utilized in the primary battery as discharge state, and the ZnO as charge state is mostly used in the secondary battery. However, because ZnO has semiconductor characteristics and poor conductivity, if the zinc anode material of nickel–zinc secondary battery is all ZnO, it will have high initial resistance and its electrode must be activated by low current charging, zinc is formed by reducing a certain proportion of ZnO, which can reduce the resistance of zinc anode, improve the charge–discharge efficiency and improve the electrochemical performance of the electrode. Therefore, the active substance of zinc anode in nickel–zinc secondary battery is mainly ZnO (charge state) with some metal Zn (discharge state) to reduce the high resistance during initial charge, and some corrosion inhibitors to improve the performance of zinc anode. Because of the solubility of ZnO in alkaline electrolyte, the dendrite morphology and passivation of zinc anode will be affected.

2.5.2.2 Zn–air battery

Zinc–air battery is a special type of fuel cell, which has been developed for more than 100 years, using oxygen in the air as its positive electrode, zinc as its negative electrode and water soluble electrolyte as its electrolyte. The main product types are divided into zinc–air primary battery and zinc–air secondary battery. The primary battery is often used in street lamps. During the operation of the direct rechargeable zinc–air secondary battery, the metal zinc electrode is dissolved or deposited, and the discharge products are dissolved in the alkaline electrolyte, the electrochemical reaction of oxygen reduction (ORR) or oxygen precipitation (OER) was carried out. The zinc–air battery chemical equation is as follows:

$$\text{Anode: } Zn + 2OH^- \rightarrow ZnO + H_2O + 2e^- \tag{2.41}$$

$$\text{Cathode: } O_2 + 2H_2O + 4e^- \rightarrow 4OH^- \tag{2.42}$$

$$\text{Totalreaction: } 2Zn + O_2 \rightarrow 2ZnO \tag{2.43}$$

Zinc electrode acts as a negative reactant for the zinc–air battery, and reacts with the positive reactant oxygen to convert chemical energy into electrical energy. Oxygen comes from the air, so the zinc–air battery should work in an open system. However, when exposed to air, the zinc–air battery metal zinc reacts with oxygen in the air even when not in working condition. So, the commercial design of the zinc–air battery requires, on the one hand, an adequate supply of oxygen at work and, on the other hand, the ability to seal off the air in non-working conditions. In addition, the problem of passivation and self-corrosion of zinc electrode is also an important aspect that restricts the performance of the zinc–air battery.

2.5.2.3 Zn–MnO_2 batteries

Zinc–manganese dioxide (Zn–MnO_2) batteries have dominated the primary battery market because of low cost, high safety, and easy manufacturing. It is highly intriguing to develop rechargeable Zn–MnO_2 batteries. Between 1986 and 1988, Yamamoto and Shoji *et al* reported rechargeable aqueous Zn–MnO_2 batteries with mild aqueous electrolytes such as aqueous $ZnSO_4$ electrolyte instead of common alkaline electrolyte, which opened the door for rechargeable aqueous ZIB based on near-neutral electrolytes.

MnO_2 is given much attention because of its high specific capacitance (308 mA h g^{-1}) and suitable discharge potential (~1.35 V). The MnO_6 octahedral units of different crystal types are connected together by edges and/or angles. Figure 2.14 displays some related details of Zn/MnO_2 systems. As depicted in figure 2.14(a), MnO_2 polymorphs could be categorized to three types, including tunnel-type structure, layered structure, and spinel type structures [128]. Most of them could be used as Zn^{2+} intercalation electrode. For β-MnO_2, which was previously considered to be detrimental to zinc interaction, the high reversible capacity of 225 mAh g^{-1} was achieved by preparing 3M $Zn(CF_3SO_3)_2$ +0.1M $Mn(CF_3SO_3)_2$ electrolyte at 0.65 °C (figure 2.14(b)) [128]. Early studies of zinc-ion chemistry have shown that Zn^{2+} is reversibly inserted into/extracted from tunnel-type α-MnO_2 in a mild $ZnSO_4$ aqueous electrolyte (figure 2.14(c)) [129], and the reaction mechanism of the battery is shown:

$$Zn^{2+} + 2e^- + 2MnO_2 \leftrightarrow ZnMn_2O_4 \tag{2.44}$$

Various Zn-ion interaction reactions and complex phase transitions of manganese oxides also have been reported, such as Zn–buserite, Zn–birnessite, and Mn^{2+} species etc.

2.5.2.4 Zn–Ag_2O batteries

Zn/Ag_2O batteries have attracted much attention due to their stable output voltage, tiny volume excellent weavability and environmentally friendliness, and they have been extensively used in consumer electronic devices, such as film cameras and

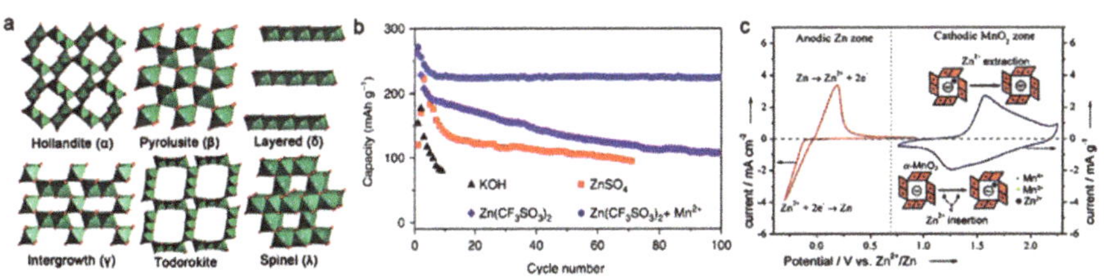

Figure 2.14. (a) Crystal structures of representative MnO_2 polymorphs. Source: [127]. Reproduced with permission of Royal Society of Chemistry. (b) Comparison of the cycling performance of Zn–MnO_2 cells with electrolytes of 45 wt% KOH, 3 M $ZnSO_4$, 3 M Zn $(CF_3SO_3)_2$, and 3 M $Zn(CF_3SO_3)_2$ + 0.1 M $Mn(CF_3SO_3)_2$. Source: [128]. Reproduced with permission of Nature Publishing Group. (c) CVs of the Zn anode (red line) and the a-MnO_2 cathode (blue line) in the $ZnSO_4$ aqueous electrolyte. Source: [129]. Reproduced with permission of Wiley-VCH.

calculators. Zn/Ag_2O batteries are the commercial rechargeable batteries in one of the highest power density products (up to 600 W kg^{-1} continuous). Furthermore, Zn/Ag_2O batteries have other advantageous properties, including high energy density (~300 Wh kg^{-1}), low self-discharge rate (~5% per month) and flat voltage during discharge process. These superior performances demonstrate that Zn/Ag_2O batteries have promising prospects for future energy storage [130]. The electrochemical reaction of $Zn–Ag_2O$ is as follows:

$$\text{Anode: } Zn + 2OH^- \rightarrow Zn(OH)_2 + 2e^- \rightarrow ZnO + H_2O + 2e^- \quad (2.45)$$

$$\text{Cathode: } Ag_2O + H_2O + 2e^- \rightarrow 2Ag + 2OH^- \quad (2.46)$$

$$\text{Total reaction: } Zn + Ag_2O \rightarrow ZnO + 2Ag \quad (2.47)$$

2.5.3 Challenge of Zn anode

2.5.3.1 The basic principle of dendrite formation

In alkaline zinc-based batteries, the formation of dendrite during the formation of zinc is a common problem in the electrodeposition process [126, 131]. Zinc has a high electrochemical activity and is thermally unstable in this medium. In alkaline electrolytes, the movement of zinc states on the zinc metal anode surface (active charge carrier) is a 'diffusion-controlled' process [132]. Initially zincate, which absorbs charge on the surface and preferentially reduces the energy-favorable charge position and forms, a small tip can cause a spike effect and attract more zinc ions to gather and form dendrites, which are gradually formed during repeated charging and discharging processes (figure 2.15(a)). This mechanism applies to neutral or weakly acidic electrolytes, but using zinc ions instead of zinc salts as charge carriers [133].

In terms of mechanism, the zinc ions adsorbed on the zinc metal anode surface diffuse two-dimensionally along the surface and gather at the preferred nucleation position to form an initial process (figure 2.15(b)), which further aggravates the typical uneven distribution of electric fields in liquid electrolytes [134]. Later

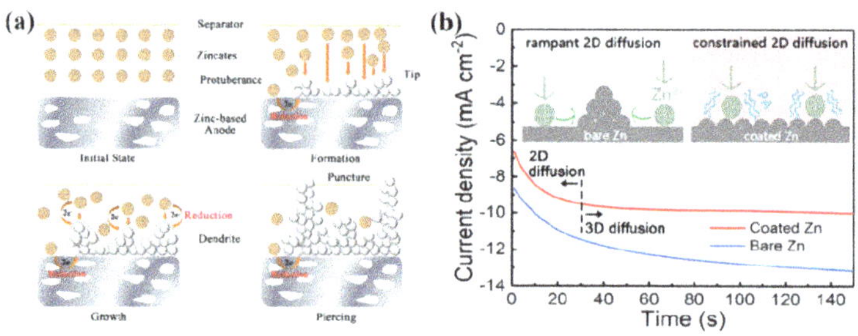

Figure 2.15. (a) Schematic diagram of the formation mechanism of zinc dendrites in alkaline electrolytes. Source: [136] Reproduced with permission of Wiley-VCH. (b) Schematic diagram of the two-dimensional diffusion of Zn^{2+} on the zinc anode surface in a weakly acidic electrolyte. Source: [137] Reproduced with permission of Royal Society of Chemistry.

carriers tend to minimize the subsistent protrusions to reduce surface energy and bare areas, leading to the growth of the protrusions and the eventual formation of dendrites [135]. Although the formation of dendrites can be effectively reduced in a mild acidic electrolyte, it remains a significant challenge, especially at high current densities. According to this basic principle, in summary, the factors leading to the formation of dendrite mainly involve inhomogeneity electric field distribution on anode surface and unlimited 2D diffusion of zinc ions.

2.5.3.2 Corrosion principle

Zinc-based battery corrosion mainly includes self-corrosion and electrochemical corrosion [138]. The former mostly exists in alkaline media, because the negative redox potential of zinc is greater than that of hydrogen, which will be studied in section 2.5.3.3. Therefore, this section will focus on the zinc anode electrochemical corrosion. The electrochemical corrosion mentioned here refers to the irreversible consumption of zinc during repeated charging and discharging, which is mainly composed of zinc dripping from the electrode into the electrolyte and inert by-products. The separation of zinc (dendrite form) from the Zn anode has been discussed in 2.5.3.1. Therefore, we will focus on inert corrosion by-products in this section.

The formation of by-products with no electrochemical activity can be observed under mild conditions, as can the formation of acidic electrolytes. We chose $ZnSO_4[Zn(OH)_2]_3xH_2O$ (ZSH) as the electrolyte of $ZnSO_4$ on behalf of expanding our discussion. ZSH is observed in both anode and cathode after cycling [138]. Figures 2.16(a) and 2.16(b) show the formation process of ZSH on the cathode surface with showing the insertion of H^+ (figure 2.16(c)). In detail, H^+ is initially inserted into the MnO_2 (cathode, material) during the first discharge phase, as H^+ concentration decreases, the OH^- concentration around the cathode is not sufficient which is beneficial to the formation of ZSH during this time. In the next stage, the

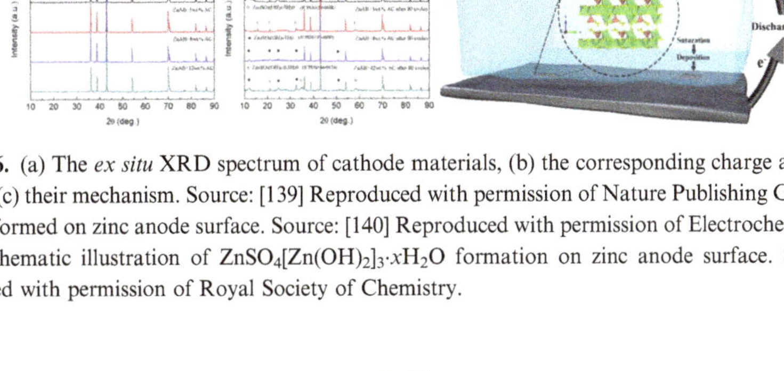

Figure 2.16. (a) The *ex situ* XRD spectrum of cathode materials, (b) the corresponding charge and discharge curve and (c) their mechanism. Source: [139] Reproduced with permission of Nature Publishing Group. (d)–(e) The ZSH formed on zinc anode surface. Source: [140] Reproduced with permission of Electrochemical Society Inc. (f) Schematic illustration of $ZnSO_4[Zn(OH)_2]_3{\cdot}xH_2O$ formation on zinc anode surface. Source: [135] Reproduced with permission of Royal Society of Chemistry.

insertion of Zn^{2+} occurs and the amount of ZSH that is formed on the surface of MnO_2 increases with the continuous decrease of H^+ concentration. This explanation also applies to other cathode materials, such as V-based materials. The ZSH formation process is described below on MnO_2:

$$H_2O \leftrightarrow H^+ + OH^- \tag{2.48}$$

$$MnO_2 + H^+ + e^- \leftrightarrow MnOOH \tag{2.49}$$

$$1/2\ Zn^{2+} + OH - +1/6ZnSO_4 + x/6H_2O \leftrightarrow 1/6\ ZnSO_4[Zn(OH)_2]_3xH_2O \tag{2.50}$$

Differing from the H^+/Zn^{2+} co-imbedding mechanism on the surface of cathode, the ZSH formation mechanism on the anode surface (figure 2.16(d) and 16(e)) is mainly attributed to the static electricity, limited solubility and interaction. During the discharge process, the zinc metal anode undergoes a conversion from Zn to Zn^{2+}, and Zn anode surface is therefore submerged by a net positive charge, which attracts SO_4^{2-} and OH^- to perform a complex reaction in the electrolyte. However, the resulting ZSH has a very limited solubility when supersaturated in the ZSH electrolyte (figure 2.16(f)). This forming process is very similar to the formation process of zinc oxide in alkaline medium. These by-products are formed due to the continual consumption of electrolyte and active Zn^{2+}, to a certain extent, which leads to low electroplating/stripping coulombic efficiency (CE) of Zn metal anode. Thus, excessive Zn is required to maintain continuous circulation, which hinders the full utilization of its theoretical capacity. In addition, the inert by-products on the electrode surface will hinder the transmission of ions, which is not conducive to the reversibility of the zinc anode.

2.5.3.3 The basic principle of hydrogen evolution

Different from lithium-ion batteries, the formation of dendrite is the main cause of battery failure [141]. Hydrogen evolution occurs at the electrode/electrolyte interface (figure 2.17(a)). The hydrogen evolution is influenced by thermodynamic and kinetic

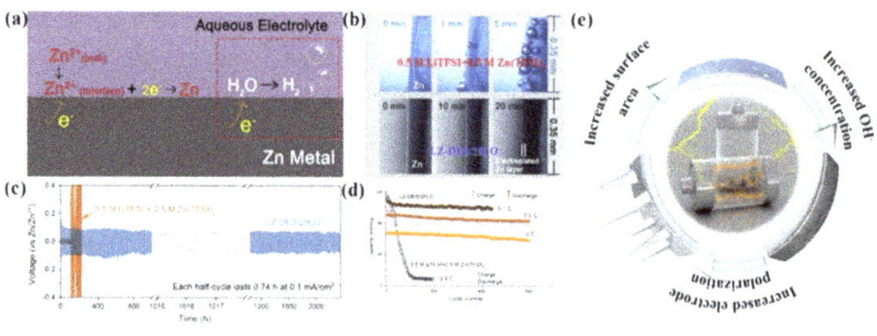

Figure 2.17. (a) Schematic illustration of hydrogen evolution. (b) *In situ* optical microscopic images of hydrogen evolution in mild acidic electrolyte. (c)–(d) Cycling performance comparison of Zn/Zn symmetrical cell and full batteries with different electrolytes. Source: [143]. Reproduced with permission of Elsevier. (e) Schematic illustration of the interaction of zinc dendrite formation, hydrogen evolution and inert corrosion by-products formation. Source: [135]. Reproduced with permission of Royal Society of Chemistry.

factors, from liquid phase to gas phase, which will increase the internal pressure, if the battery is a sealed system, the battery will be damaged due to internal expansion. The deposition of zinc is inevitably affected by competitive hydrogen gas release, which not only reduces CE, but also leads to local pH fluctuation due to the increase of OH^- concentration. And the generated OH^- is an important source of ZSH formation on the anode surface.

The standard reduction potential of ZnO/Zn (−1.26 V vs. SHE) is lower than that of hydrogen, precipitation potential in alkaline medium (−0.83 V vs. SHE), and their reactions formulas are displayed are as follows:

$$ZnO + 2e^- + H_2O \leftrightarrow Zn + 2OH^-(-1.26V) \tag{2.51}$$

$$2H_2O + 2e^- \leftrightarrow H_2 \uparrow + 2OH^-(-0.83V) \tag{2.52}$$

According to this principle, the higher the standard reduction potential, it is easier to acquire electrons in the oxidized state of matter, obviously, hydrogen is precipitated prior to hydrogen, and the thermodynamic reduction of ZnO is a thermodynamically preferred, alkaline electrolyte. Such severe hydrogen release limits the practical use of hydrogen, an alkaline zinc-based battery, because it not only threatens cycle lifespan, but also shelf life and resilience.

By contrast, the standard reduction potential of Zn/Zn^{2+} (−0.76 V vs. SHE) is higher than that of Zn/Zn^{2+}. The value of hydrogen evolution in a neutral medium, which means that hydrogen evolution does not preempt the thermodynamic reduction of zinc ions. However, in slightly acidic electrolyte, the standard reduction potential of Zn/Zn^{2+} lags behind the hydrogen evolution potential (0V vs. SHE) and appears to have a significant hydrogen release trend. Their reaction formula is as follows:

$$Zn^{2+} + 2e^- \leftrightarrow Zn(-0.76V) \tag{2.53}$$

$$2H + +2e^- \leftrightarrow H_2 \uparrow (0V) \tag{2.54}$$

In fact, the hydrogen release is not as severe as these values suggest because of the high zinc content. The overpotential of hydrogen evolution in water environment can be explained by the Tafel equation:

$$\eta = b \log i + a \tag{2.55}$$

where η is the hydrogen release overpotential, i is the current density, b is the Tafel slope constant and a is the overpotential when current density i is equal to the unit current density.

In formula (2.55), the b constants of different metals are almost the same (~0.12 V), so the value of the hydrogen release overpotential is mainly determined by the a constant. Zinc is a metal with high hydrogen overpotential, with a relatively high a constant. At this high potential, the release of hydrogen is consistent with the Pourbaix diagram of zinc in aqueous solution and leads to the inhibition of hydrogen precipitation on the zinc surface. In practice, however, hydrogen overpotential is also affected by kinetic factors, including electrode surface, operating temperature, zinc concentration, etc. Thus, hydrogen can be obtained in slightly acidic electrolytes under certain conditions (figure 2.17(b)), resulting in poor recyclability (figure 2.17(c) and 2.17(d)).

2.5.3.4 The interaction between above factors

In fact, the different issues discussed above are not independent, but are of great significance interacting with each other [142]. Their interactions are summarized and shown in figure 2.4(e). In detail, the formation of dendrites will lead to an increase in the surface area of the anode, thus accelerating the anodic oxidation and hydrogen evolution. The escape of hydrogen from the surface of the anode can lead to anodic oxidation, and the local pH changes due to the increase of the concentration of OH^-. These produced OH^- will participate in the reactions and formation an electrochemical inert corrosion by-products, which is deposited on the surface of the zinc metal anode. Inert corrosion by-products on the zinc anode surface may lead to inhomogeneity surface and enlarged overpotential, which in turn promotes zinc dendrite formation. Therefore, in practice, when we aim to solve one of the above problems, we may alleviate one or more of the other problems indirectly. Instead, we adopt solutions that suppress one problem and, in some cases, may exacerbate another.

2.6 Outlook and summary

Na, K, Mg, Al, and Zn metal batteries are rich in resources, have high energy density, and have electrochemical characteristics similar to Li metal. They have shown good prospects in energy storage devices and have received widespread attention. Here we discussed the problems of dendritic growth and side reactions in metal anodes, and summarized strategies such as alloy anodes, 3D structures, solid electrolytes, artificial SEI and electrolyte engineering. Although tremendous progress has been made in recent years, there is still a long way to go before the application of metal anodes. There are still shortcomings in the development of metal batteries under extreme conditions, the reaction mechanism of the interface between the electrode and the electrolyte, the establishment of the theoretical model of the interface, and the non-destructive *in situ* characterization analysis. Therefore, the development and application of metal batteries need to be continuously explored and optimized to reach a higher level.

References

[1] Zhao C, Lu Y, Yue J, Pan D, Qi Y, Hu Y-S and Chen L 2018 *J. Energy Chem.* **27** 1584–96

[2] Lee B, Paek E, Mitlin D and Lee S W 2019 *Chem. Rev.* **119** 5416–60

[3] Kim Y, Kim J K, Vaalma C, Bae G H, Kim G T, Passerini S and Kim Y 2018 *Nat. Rev. Mater.* **3** 564–71

[4] Wang Z, Liu Y, Wu Z, Guan G, Zhang D, Zheng H, Xu S, Liu S and Hao X 2017 *Nanoscale* **9** 823–31
Wang H, Matios E, Luo J and Li W 2020 *Chem. Soc. Rev.* **49** 3783-805

[5] Peters J, Buchholz D, Passerini S and Weil M 2016 *Energy Environ. Sci.* **9** 1744–51

[6] Vesborg P C K and Jaramillo T F 2012 *RSC Adv.* **2** 7933

[7] Hwang J Y, Myung S T and Sun Y K 2017 *Chem. Soc. Rev.* **46** 3529–614
Yang Z, Zhang J, Kintner-Meyer M C, Lu X, Choi D, Lemmon J P and Liu J 2011 *Chem. Rev.* **111** 3577–613
Xiang X, Zhang K and Chen J 2015 *Adv. Mater.* **27** 5343–64
Islam M S and Fisher C A 2014 *Chem. Soc. Rev.* **43** 185–204

[8] Yabuuchi N, Kubota K, Dahbi M and Komaba S 2014 *Chem. Rev.* **114** 11636–82
Luo W, Shen F, Bommier C, Zhu H, Ji X and Hu L 2016 *Acc Chem. Res.* **49** 231–40
Bauer A, Song J, Vail S, Pan W, Barker J and Lu Y 2018 *Adv. Energy Mater.* **8** 1702869

[9] Yabuuchi N, Kajiyama M, Iwatate J, Nishikawa H, Hitomi S, Okuyama R, Usui R, Yamada Y and Komaba S 2012 *Nat. Mater.* **11** 512–17
Kundu D, Talaie E, Duffort V and Nazar L F 2015 *Angew Chem Int. Ed. Engl.* **54** 3431–48

[10] Kubota K and Komaba S 2015 *J. Electrochem. Soc.* **162** A2538-A2550

[11] Lee B, Kim M, Kim S, Nanda J, Kwon S J, Jang H D, Mitlin D and Lee S W 2020 *Adv. Energy Mater.* **10** 1903280

[12] Kim H, Kim H, Ding Z, Lee M H, Lim K, Yoon G and Kang K 2016 *Adv. Energy Mater.* **6** 1600943
Liang Y, Lai W H, Miao Z and Chou S L 2018 *Small* **14** 1702514

[13] Deng J, Luo W-B, Chou S-L, Liu H-K and Dou S-X 2018 *Adv. Energy Mater.* **8** 1701428
Slater M D, Kim D, Lee E and Johnson C S 2013 *Adv. Funct. Mater.* **23** 947–58
Song J H, Xiao B W, Lin Y H, Xu K and Li X L 2018 *Adv. Energy Mater.* **8** 1703082

[14] Luo W and Hu L 2015 *ACS Cent Sci.* **1** 420–22

[15] Yadegari H, Sun Q and Sun X 2016 *Adv. Mater.* **28** 7065–93

[16] Peled E, Golodnitsky D, Mazor H, Goor M and Avshalomov S 2011 *J. Power Sources* **196** 6835–40

[17] Sun Q, Yang Y and Fu Z-W 2012 *Electrochem. Commun.* **16** 22–5

[18] Zhao Y, Adair K R and Sun X 2018 *Energy Environ. Sci.* **11** 2673–95

[19] Zhao N, Li C and Guo X 2014 *Phys. Chem. Chem. Phys.* **16** 15646–52
Medenbach L, Bender C L, Haas R, Mogwitz B, Pompe C, Adelhelm P, Schröder D and Janek J 2017 *Energy Technol.* **5** 2265–74

[20] Bi X, Ren X, Huang Z, Yu M, Kreidler E and Wu Y 2015 *Chem. Commun (Camb)* **51** 7665–68

[21] Ottakam Thotiyl M M, Freunberger S A, Peng Z and Bruce P G 2013 *J. Am. Chem. Soc.* **135** 494–500
LiuG T, Kim M T, Casford C P and Grey 2016 *J. Phys. Chem. Lett.* **7** 4841–46

[22] Yadegari H, Li Y, Banis M N, Li X, Wang B, Sun Q, Li R, Sham T-K, Cui X and Sun X 2014 *Energy Environ. Sci.* **7** 3747–57

[23] Yadegari H *et al* 2015 *Chem. Mater.* **27** 3040–47

[24] Song K, Agyeman D A, Park M, Yang J and Kang Y M 2017 *Adv. Mater.* **29** 1606572

[25] Su H Y, Gorlin Y, Man I C, Calle-Vallejo F, Norskov J K, Jaramillo T F and Rossmeisl J 2012 *Phys. Chem. Chem. Phys.* **14** 14010–22

[26] Bieker G, Winter M and Bieker P 2015 *Phys. Chem. Chem. Phys.* **17** 8670–79

[27] Cheng X-B and Zhang Q 2015 *J. Mater. Chem.* A **3** 7207–9

[28] Lutz L, Alves Dalla Corte D, Tang M, Salager E, Deschamps M, Grimaud A, Johnson L, Bruce P G and Tarascon J-M 2017 *Chem. Mater.* **29** 6066–75

[29] Lee Y, Lee J, Lee J, Kim K, Cha A, Kang S, Wi T, Kang S J, Lee H W and Choi N S 2018 *ACS Appl. Mater. Interfaces* **10** 15270–80
Han M, Zhu C, Ma T, Pan Z, Tao Z and Chen J 2018 *Chem Commun (Camb)* **54** 2381–84
[30] Wu S, Qiao Y, Jiang K, He Y, Guo S and Zhou H 2018 *Adv. Funct. Mater.* **28** 1706374
[31] Wang H, Wang C, Matios E and Li W 2018 *Angew. Chem. Int. Ed. Engl.* **57** 7734–37
[32] Shi Q, Zhong Y, Wu M, Wang H and Wang H 2018 *Angew. Chem. Int. Ed. Engl.* **57** 9069–72
[33] Choudhury S *et al* 2017 *Nat. Commun.* **8** 898
[34] Zhu M, Wang G, Liu X, Guo B, Xu G, Huang Z, Wu M, Liu H K, Dou S X and Wu C 2020 *Angew. Chem. Int. Ed. Engl.* **59** 6596–600
[35] Meng X, Yang X Q and Sun X 2012 *Adv. Mater.* **24** 3589–615
Zhao Y, Goncharova L V, Lushington A, Sun Q, Yadegari H, Wang B, Xiao W, Li R and Sun X 2017 *Adv. Mater.* **29** 1606663
[36] Li K *et al* 2019 *Nat. Commun.* **10** 725
Shang S L, Yu Z, Wang Y, Wang D and Liu Z K 2017 *ACS Appl Mater Interfaces* **9** 16261–69
[37] Yao M *et al* 2019 *ACS Appl. Energy Mater.* **2** 2743–50
[38] Chi X, Hao F, Zhang J, Wu X, Zhang Y, Gheytani S, Wen Z and Yao Y 2019 *Nano Energy* **62** 718–24
[39] Zheng J, Chen S, Zhao W, Song J, Engelhard M H and Zhang J-G 2018 *ACS Energy Lett.* **3** 315–21
Geng C, Buchholz D, Kim G T, Carvalho D V, Zhang H, Chagas L G and Passerini S 2018 *Small Methods* **3** 1800208
Liu S, Tang S, Zhang X, Wang A, Yang Q H and Luo J 2017 *Nano Lett.* **17** 5862–68
Wu F, Zhou J, Luo R, Huang Y, Mei Y, Xie M and Chen R 2019 *Energy Storage Mater.* **22** 376–83
[40] Luo L, Liu T, Zhang S, Ke B, Yu L, Hussain S and Lin L 2017 *Ceram. Int.* **43** 5095–101
[41] Sun J, Guo C, Cai Y, Li J, Sun X, Shi W, Ai S, Chen C and Jiang F 2019 *Electrochim. Acta* **309** 18–24
[42] Hu X F, Li Z F, Zhao Y R, Sun J C, Zhao Q, Wang J B, Tao Z L and Chen J 2017 *Sci. Adv.* **3** 1602396
Wang A, Hu X, Tang H, Zhang C, Liu S, Yang Y W, Yang Q H and Luo J 2017 *Angew. Chem. Int. Ed. Engl.* **56** 11921–26
Luo W, Zhang Y, Xu S, Dai J, Hitz E, Li Y, Yang C, Chen C, Liu B and Hu L 2017 *Nano Lett.* **17** 3792–97
[43] Chen H, Liu P, Liu J, Feng X and Zhou S 2021 *J. Catal.* **394** 397–405
[44] Hu X, Sun J, Li Z, Zhao Q, Chen C and Chen J 2016 *Angew. Chem. Int. Ed. Engl.* **55** 6482–86
Ma W, Liu X, Li C, Yin H, Xi W, Liu R, He G, Zhao X, Luo J and Ding Y 2018 *Adv. Mater.* **30** e1801152
Kim J, Seong A, Yang Y, Joo S, Kim C, Jeon H D, Dai L and Kim G 2021 *Nano Energy* **82** 105741
[45] Mu X, Pan H, He P and Zhou H 2020 *Adv. Mater.* **32** 1903790
[46] Zheng Z, Wu C, Gu Q, Konstantinov K and Wang J 2021 *Energy Environ. Mater.* **4** 158–77
[47] Lu Y, Cai Y, Zhang Q, Liu L, Niu Z and Chen J 2019 *Chem. Sci.* **10** 4306–12

[48] Wang X, Zhang X, Lu Y, Yan Z, Tao Z, Jia D and Chen J 2018 *ChemElectroChem.* **5** 3628–32
[49] Guo L, Li B, Thirumal V and Song J 2019 *Chem. Commun (Camb)* **55** 7946–49
Fang C, Luo J, Jin C, Yuan H, Sheng O, Huang H, Gan Y, Xia Y, Liang C, Zhang J, Zhang W and Tao X 2018 *ACS Appl. Mater. Interfaces* **10** 17240–48
[50] Kumar D, Rajouria S K, Kuhar S B and Kanchan D K 2017 *Solid State Ionics* **312** 8–16
[51] Nikiforidis G, van de Sanden M C M and Tsampas M N 2019 *RSC Adv.* **9** 5649–73
[52] Wang Y X, Zhang B, Lai W, Xu Y, Chou S L, Liu H K and Dou S X 2017 *Adv. Energy Mater.* **7** 1602829
Abraham K, Rauh R and Brummer S 1978 *Electrochim. Acta* **23** 501–7
[53] Yu X and Manthiram A 2014 *J. Phys. Chem. Lett.* **5** 1943–47
[54] Zhou W, Li Y, Xin S and Goodenough J B 2017 *ACS Cent Sci.* **3** 52–7
Adelhelm P, Hartmann P, Bender L C, Busche M, Eufinger C and Janek J 2015 *Beilstein J. Nanotechnol.* **6** 1016–55
[55] Lee D-J, Park J-W, Hasa I, Sun Y-K, Scrosati B and Hassoun J 2013 *J. Mater. Chem.* A **1** 5256
[56] Seh Z W, Sun J, Sun Y and Cui Y 2015 *ACS Cent. Sci.* **1** 449-55
[57] La Rosa D, Monforte G, D'Urso C, Baglio V, Antonucci V and Arico A S 2010 *ChemSusChem.* **3** 1390–97
Lu X, Evans J R G and Heavens S N 2012 *J. Eur. Ceram. Soc.* **32** 4219–28
Zhu C, Xue J and Ji G 2015 *Mater. Sci. Semiconductor Process.* **31** 487–92
Yi E, Temeche E and Laine R M 2018 *J. Mater. Chem. A* **6** 12411–19
[58] Yu X and Manthiram A 2014 *J. Phys. Chem.* C **118** 22952–59
[59] Sun Q, He B, Zhang X Q and Lu A H 2015 *ACS Nano* **9** 8504–13
Suzuki K, Mashimo N, Ikeda Y, Yokoi T, Hirayama M and Kanno R 2018 *ACS Appl. Energy Mater.* **1** 2373–77
Fan X, Yue J, Han F, Chen J, Deng T, Zhou X, Hou S and Wang C 2018 *ACS Nano* **12** 3360–68
Hu J, Zhong H, Yan X and Zhang L 2018 *Appl. Surf. Sci.* **457** 705–11
[60] Cheng X B, Zhang R, Zhao C Z and Zhang Q 2017 *Chem. Rev.* **117** 10403–73
[61] Hwang J-Y, Myung S-T and Sun Y-K 2018 *Adv. Funct. Mater.* **28** 1802938
[62] Hamon Y, Brousse T, Jousse F, Topart P, Buvat P and Schleich D M 2001 *J. Power Sources* **97-8** 185–87
[63] Ding J, Zhang H, Fan W, Zhong C, Hu W and Mitlin D 2020 *Adv. Mater.* **32** e1908007
Komaba S, Hasegawa T, Dahbi M and Kubota K 2015 *Electrochem. Commun.* **60** 172–75
Kubota K, Dahbi M, Hosaka T, Kumakura S and Komaba S 2018 *Chem. Rec.* **18** 459–79
Pham T A, Kweon K E, Samanta A, Lordi V and Pask J E 2017 *J. Phys. Chem. C* **121** 21913–20
[64] Chazalviel J 1990 *Phys. Rev.* A **42** 7355–67
Brissot C, Rosso M, Chazalviel J N and Lascaud S 1999 *J. Power Sources* **81** 925–29
[65] Ju Z, Zhang S, Xing Z, Zhuang Q, Qiang Y and Qian Y 2016 *ACS Appl. Mater. Interfaces* **8** 20682–90
Share K, Cohn P A, Carter R, Rogers B and Pint C L 2016 *ACS Nano* **10** 9738–44
Chen C, Wang Z, Zhang B, Miao L, Cai J, Peng L, Huang Y, Jiang J, Huang Y, Zhang L and Xie J 2017 *Energy Storage Mater.* **8** 161–68

Qi X, Huang K, Wu X, Zhao W, Wang H, Zhuang Q and Ju Z 2018 *Carbon* **131** 79–85
Ma G, Huang K, Ma J-S, Ju Z, Xing Z and Zhuang Q-C 2017 *J. Mater. Chem. A* **5** 7854–61
Adams R A, Syu J M, Zhao Y, Lo C T, Varma A and Pol V G 2017 *ACS Appl Mater Interfaces* **9** 17872–81
Wang X, Sun G, Routh P, Kim H D, Huang W and Chen P 2014 *Chem Soc Rev* **43** 7067–98
[66] Kishore B, G V and Munichandraiah N 2016 *J. Electrochem. Soc.* **163** A2551–554
Han J, Xu M, Niu Y, Li G N, Wang M, Zhang Y, Jia M and Li C M 2016 *Chem Commun (Camb)* **52** 11274–76
[67] Deng Q, Pei J, Fan C, Ma J, Cao B, Li C, Jin Y, Wang L and Li J 2017 *Nano Energy* **33** 350–55
Lei K, Li F, Mu C, Wang J, Zhao Q, Chen C and Chen J 2017 *Energy Environ. Sci.* **10** 552–57
[68] Liu P, Wang Y, Gu Q, Nanda J, Watt J and Mitlin D 2020 *Adv. Mater.* **32** e1906735
[69] Hwang J-Y, Kim H M and Sun Y-K 2018 *J. Mater. Chem.* A **6** 14587–93
Lu X, Bowden M E, Sprenkle V L and Liu J 2015 *Adv Mater* **27** 5915–22
[70] Ma R, Fan L, Wang J and Lu B 2019 *Electrochim. Acta* **293** 191–98
[71] Xiong P, Han X, Zhao X, Bai P, Liu Y, Sun J and Xu Y 2019 *ACS Nano* **13** 2536–43
[72] Elazari R, Salitra G, Garsuch A, Panchenko A and Aurbach D 2011 *Adv. Mater.* **23** 5641–44
Kim J, Lee D-J, Jung H-G, Sun Y-K, Hassoun J and Scrosati B 2013 *Adv. Funct. Mater.* **23** 1076–80
Zou Q, Liang Z, Du G Y, Liu C Y, Li E Y and Lu Y C 2018 *J. Am. Chem. Soc.* **140** 10740–48
[73] Lai N-C, Cong G and Lu Y-C 2019 *J. Mater. Chem.* A **7** 20584–89
Ma S, Zuo P, Zhang H, Yu Z, Cui C, He M and Yin G 2019 *ChemCommun. (Camb)* **55** 5267–70
[74] Gu S, Xiao N, Wu F, Bai Y, Wu C and Wu Y 2018 *ACS Energy Lett.* **3** 2858–64
[75] Wang L, Bao J, Liu Q and Sun C-F 2019 *Energy Storage Mater.* **18** 470–75
Hwang J-Y, Kim H M, Yoon C S and Sun Y-K 2018 *ACS Energy Lett* **3** 540–41
Zhao Q, Hu Y, Zhang K and Chen J 2014 *Inorg. Chem.* **53** 9000–5
[76] Yu X and Manthiram A 2018 *Energy Storage Mater.* **15** 368–73
Liu D-H, Li W-H, Liang H-J, Lue H-Y, Guo J-Z, Wang J and Wu X-L 2018 *J. Mater. Chem. A* **6** 15797–806
[77] Yuan X, Zhu B, Feng J, Wang C, Cai X, Qiao K and Qin R 2020 *Small* **16** e2003386
[78] Tang X, Zhou D, Li P, Guo X, Sun B, Liu H, Yan K, Gogotsi Y and Wang G 2020 *Adv. Mater.* **32** e1906739
[79] Yao Y, Xu R, Chen M, Cheng X, Zeng S, Li D, Zhou X, Wu X and Yu Y 2019 *ACS Nano* **13** 4695–704
[80] Liu Q, Deng W, Pan Y and Sun C F 2020 *Chem. Sci.* **11** 6045–52
[81] Jin J, Tian X, Srikanth N, Kong L B and Zhou K 2017 *J. Mater. Chem.* A **5** 10110–26
[82] Zeng L, Zeng W, Jiang Y, Wei X, Li W, Yang C, Zhu Y and Yu Y 2015 *Adv. Energy Mater.* **5** 1401377
[83] Liu Y, Tai Z, Zhang Q, Wang H, Pang W K, Liu H K, Konstantinov K and Guo Z 2017 *Nano Energy* **35** 36–43
[84] Yao Y, Chen M, Xu R, Zeng S, Yang H, Ye S, Liu F, Wu X and Yu Y 2018 *Adv. Mater.* **30** e1805234
[85] Xu R *et al* 2020 *Adv. Mater.* **32** 2003879
[86] Shiraz M H A, Zhao P and Liu J 2020 *J. Power Sources* **453** 227855
[87] Xiao N, Zheng J, Gourdin G, Schkeryantz L and Wu Y 2019 *ACS Appl. Mater. Interfaces* **11** 16571–77

[88] Huang X, Sun J, Wang L, Tong X, Dou S X and Wang Z M 2021 *Small* **17** e2004369
[89] He J, Lv W, Chen Y, Xiong J, Wen K, Xu C, Zhang W, Li Y, Qin W and He W 2017 *J. Power Sources* **363** 103–9
Wu T, Ding Z, Jing M, Zou G, Hou H, Tian Y, Jiang Y, Hong W and Ji X 2019 *Adv. Funct. Mater.* **29** 1809014
Li F and Zhou Z 2018 *Small* **14** 1702961
Zhang W, Wang H, Zhang N, Liu H, Chen Z, Zhang L, Guo S, Li D and Xu J 2019 *ACS Appl. Mater. Interfaces* **11** 29807–13
[90] Bella F, De Luca S, Fagiolari L, Versaci D, Amici J, Francia C and Bodoardo S 2021 *Nanomaterials* **11** 810
[91] Song M J, Niu J Z, Gao H, Kou T Y, Wang Z K and Zhang Z H 2020 *J. Mater. Chem.* A **8** 13572–84
Saha P, Datta M K, Velikokhatnyi O I, Manivannan A, Alman D and Kumta P N 2014 *Prog. Mater. Sci.* **66** 1–86
[92] Jia D, Liu F, Yang D Z and Wang W 2020 *Int. J. Electrochem. Sci.* **15** 10584–615
[93] Niu J Z, Zhang Z H and Aurbach D 2020 *Adv. Energy Mater.* **10** 2000697
[94] Li Z, Han L, Wang Y F, Li X Y, Lu J L and Hu X W 2019 *Small* **15** 1900105
[95] Wu D H, Yang B C, Zhang S R, Ruckenstein E and Chen H Y 2021 *J. Colloid Interface Sci.* **597** 401–8
[96] Mukherjee A, Taragin S, Aviv H, Perelshtein I and Noked M 2020 *Adv. Funct. Mater.* **30** 2003518
[97] Tran N A, Thanh N D V and Le M L P 2011 *Chem.-A Eur. J.* **27** 9198–217
Matsui M 2011 *J. Power Sources* **196** 7048–55
[98] Zhang E, Xing Z, Wang J, Ju Z and Qian Y 2012 *RSC Adv.* **2** 6748–51
[99] Wilcock W S D and Kauffman P C 1997 *J. Power Sources* **66** 71–5
Zhao H, Bian P and Ju D 2009 *J. Environ. Sci. (China)* **21 Suppl 1** S88–91
[100] Aurbach D, Lu Z, Schechter A, Gofer Y, Gizbar H, Turgeman R, Cohen Y, Moshkovich M and Levi E 2000 *Nature* **407** 724–27
[101] Tian Y S *et al* 2021 *Chem. Rev.* **121** 1623–69
[102] Guo Q, Zeng W, Liu S L, Li Y Q, Xu J Y, Wang J X and Wang Y 2021 *Rare Met.* **40** 290–308
[103] Aurbach D, Weissman I, Gofer Y and Levi E 2003 *Chem. Rec.* **3** 61–73
[104] Gregory T D, Hoffman R J and Winterton R C 1990 *J. Electrochem. Soc.* **137** 775–80
[105] Zhou W J, Zhang M, Kong X Y, Huang W W and Zhang Q C *Adv. Sci.* **8** 2004490
Muldoon J, Bucur C B, Oliver A G, Sugimoto T, Matsui M, Kim H S, Allred G D, Zajicek J and Kotani Y 2012 *Energy Environ. Sci.* **5** 5941–50
[106] Kumar G G and Munichandraiah N 1999 *Electrochim. Acta* **44** 2663–66
[107] Yoshimoto N, Yakushiji S, Ishikawa M and Morita M 2002 *Solid State Ionics* **152** 259–66
[108] Chusid O, Gofer Y, Gizbar H, Vestfrid Y, Levi E, Aurbach D and Riech I 2003 *Adv. Mater.* **15** 627–30
[109] Li C, Cheng F, Ji W, Tao Z and Chen J 2009 *Nano Res.* **2** 713–21
[110] Nguyen D T and Song S W 2017 *J. Power Sources* **368** 11–7
[111] Zhang J, Guan X, Lv R, Wang D, Liu P and Luo J 2020 *Energy Storage Mater.* **26** 408–13
[112] Pei C Y, Xiong F Y, Yin Y M, Liu Z, Tang H, Sun R M, An Q Y and Mai L Q 2021 *Small* **17** 2004108
[113] Mohtadi R, Tutusaus O, Arthur T S, Zhao-Karger Z and Fichtner M 2021 *Joule* **5** 581–617
[114] Liang Z M and Ban C M 2021 *Angew. Chem. Int. Ed.* **60** 11036–47

[115] Tarascon J M and Armand M 2001 *Nature* **414** 359–67
Etacheri V, Marom R, Elazari R, Salitra G and Aurbach D 2011 *Energy Environ. Sci.* **4** 3243–62
[116] Yabuuchi N, Kubota K, Dahbi M and Komaba S 2014 *Chem. Rev.* **114** 11636–82
Palomares V, Serras P, Villaluenga I, Hueso B K, Carretero-Gonzalez J and Rojo T 2012 *Energy Environ. Sci.* **5** 5884–901
[117] Eftekhari A and Corrochano P 2017 *Sustain. Energy Fuels* **1** 1246–64
[118] Kraychyk K V, Wang S, Piveteau L and Koyalenko M V 2017 *Chem. Mater.* **29** 4484–92
[119] Lin M-C *et al* 2015 *Nature* **520** 325
[120] Revel R, Audichon T and Gonzalez S 2014 *J. Power Sources* **272** 415–21
[121] Munoz-Torrero D, Palma J, Marcilla R and Ventosa E 2019 *Dalton Trans.* **48** 9906–11
[122] Wang S, Kravchyk K V, Krumeich F and Kovalenko M V 2017 *ACS Appl. Mater. Interfaces* **9** 28478–85
[123] Pan C-J, Yuan C, Zhu G, Zhang Q, Huang C-J, Lin M-C, Angell M, Hwang B-J, Kaghazchi P and Dai H 2018 *Proc. of the National Academy of Sciences of the United States of America* **115** 5670–675
[124] Li M, Li Z, Wang X, Meng J, Liu X, Wu B, Han C and Mai L 2021 *Energy Environ. Sci.* **14** 3796–839
[125] Zhang Q, Luan J, Tang Y, Ji X and Wang H 2020 *Angew. Chem. Int. Ed. Engl.* **59** 13180–91
[126] Du W, Ang E H, Yang Y, Zhang Y, Ye M and Li C C 2020 *Energy Environ. Sci.* **13** 3330–60
[127] Zhang N, Chen X, Yu M, Niu Z, Cheng F and Chen J 2020 *Chem. Soc. Rev.* **49** 4203–19
[128] Zhang N, Cheng F, Liu J, Wang L, Long X, Liu X, Li F and Chen J 2017 *Nat. Commun.* **8** 405
[129] Xu C, Li B, Du H and Kang F 2012 *Angew. Chem. Int. Ed. Engl.* **51** 933–35
[130] Ozgit D, Hiralal P and Amaratunga G A J 2014 *ACS Appl. Mater. Interfaces* **6** 20752–57
[131] Tang B, Shan L, Liang S and Zhou J 2019 *Energy Environ. Sci.* **12** 3288–304
Wang F-H and Liu H-B 2019 *Chinese J. Inorg. Chem.* **35** 1999–2012
[132] Fan L and Li X 2018 *Nano Energy* **53** 630–42
[133] Han C, Li W, Liu H K, Dou S and Wang J 2020 *Nano Energy* **74** 104880
[134] Jia H, Wang Z, Tawiah B, Wang Y, Chan C-Y, Fei B and Pan F 2020 *Nano Energy* **70** 104523
[135] Li C, Xie X, Liang S and Zhou J 2020 *Energy Environ. Mater.* **3** 146–59
[136] Lu W, Xie C, Zhang H and Li X 2018 *ChemSusChem* **11** 3996–4006
[137] Zhao Z, Zhao J, Hu Z, Li J, Li J, Zhang Y, Wang C and Cui G 2019 *Energy Environ. Sci.* **12** 1938–49
[138] Ma L and Zhi C 2021 *Electrochem. Commun.* **122** 106898
Chai Y J, Du Y H, Li L and Wang N 2020 *Nanotechnology* **31** 215402
[139] Huang J, Wang Z, Hou M, Dong X, Liu Y, Wang Y and Xia Y 2018 *Nat. Commun.* **9** 2906
[140] Li H, Xu C, Han C, Chen Y, Wei C, Li B and Kang F 2015 *J. Electrochem. Soc.* **162** A1439-A1444
[141] Wu H, Jia H, Wang C, Zhang J G and Xu W 2020 *Adv. Energy Mater.* **11** 2003092
Pu X, Jiang B, Wang X, Liu W, Dong L, Kang F and Xu C 2020 *Nanomicro. Lett.* **12** 152
[142] Shin J, Lee J, Park Y and Choi J W 2020 *Chem. Sci.* **11** 2028–44
[143] Zhao J, Zhang J, Yang W, Chen B, Zhao Z, Qiu H, Dong S, Zhou X, Cui G and Chen L 2019 *Nano Energy* **57** 625–34

IOP Publishing

Metal Electrodes for Battery Technologies

Yurong Ren and Jianmin Ma

Chapter 3

Germanium as the negative electrode for various batteries

Jun Wang

In current years, Ge has been considered to be the greatest anode potential candidate for Li-ion batteries, meanwhile its potential application in Na and K-ion batteries is also approaching, owing to the superior intrinsic physical and chemical merits for favorable electrochemical behaviors. Although many advances for Ge anodes had been reported, their application is still largely limited by a series of several critical challenges, including obvious volume changes, poor kinetics and unstable solid electrolyte interface (SEI) layers during cycling. This chapter offers a brief review on the latest progresses of Ge-based materials as anodes for Li, Na and K storage. The state-of-the-art strategies to obtain improved electrochemical properties are listed. Desired energy storage mechanisms are also deeply discussed, mainly related to phase transform, ionic/electronic transport kinetics and specific chemical reactions, etc. Finally, the perspective of future research of Ge-based anodes in Li, Na and K-ion batteries is provided.

3.1 Ge properties

Although Ge element is widely distributed all over the Earth, most of it exists naturally in various minerals with low concentrations, leading to hard discovering and producing to fulfill its industrial use. Since the first study on Ge-based materials as semiconductors in the mid-1940s, they have been intensively investigated and employed in research fields and practical application. Interestingly, Ge features many unique advantages, including relatively high electron mobility, high frequency and excellent mechanical strength. These endow it superior physical and chemical behaviors, which are beneficial for serving in the high frequency field, far-infrared and aerospace electronics, fiber-optic, infrared optics, photoluminescence, solar cells, hydrogen-storage, superconductors, batteries, and so on [1–3].

doi:10.1088/978-0-7503-4871-3ch3

The relative atomic mass of Ge is 72.64 with a density of ~5.5 g cm^{-3}, and it is a face centered cubic crystal with lattice parameters of a = b = c = 565.75 pm, $\alpha = \beta = \gamma = 90°$. Acting as anode materials for lithium-ion batteries (LIBs), sodium-ion batteries (SIBs) and potassium-ion batteries (PIBs), the electrochemical properties of Ge have been listed in table 3.1. Ge could react with Li to form $Li_{22}Ge_5$ and $Li_{15}Ge_4$ with capacities of 1623 [4] and 1384 mAh g^{-1} [5], respectively. In contrast, NaGe presents a theoretical capacity of 369 mAh g^{-1}, while the theoretical capacity of potassium insertion in Ge is 369 mAh g^{-1}. However, it is found that Ge anodes would show obvious volume changes during cycling.

As can be seen clearly in figure 3.1, the theoretical capacities of Ge apparently outperform those of Sn and C in terms of theoretical lithiation, sodiation and potassiation capacities. Although they are lower than those of Si, the volume changes of Ge are actually smaller, and the Li^+ and e^- transports are faster, resulting in better electrochemical kinetics in LIBs, SIBs and PIBs [10]. Because of the multiple merits, Ge-based materials are considered as ideal materials to be used as anodes for these rechargeable alkali metal-ion batteries. To meet the practical battery requirements, their electrochemical performance needs to be improved, and the progresses in recent years are summarized and presented in the following discussions.

3.1.1 Ge anode for Li-ion batteries

3.1.1.1 Ge nanostructures

Group IV elements are considered to be alternative materials for the next generation of commercial for LIBs anodes. Among them, Si features the highest theoretical capacity (3579 mA h g^{-1}), low working potential, environmental friendliness, abundant resources and low cost. However, for practical applications, due to the poor electrical conductivity and the low Li^+ diffusivity, the low initial columbic efficiency, poor capacity and rate performance limit its commercialization. In addition, Ge anodes also deliver high-rate performance and cycle performance, compared with Si counterpart, and better electrical conductivity and ion conductivity make them considered to hold greater application potential than Si anodes for LIBs [14].

Table 3.1. Electrochemical properties of Ge as anodes for LIBs, SIBs and PIBs. Source: [6–9]. Reproduced with permission of Royal Society of Chemistry, Elsevier, American Chemical Society, John Wiley and Sons Ltd.

Alkali	Stoichiometry final product	Theoretical capacity (mAh g^{-1})	Approximated volume change (%)
Li	$Li_{22}Ge_5$	1623	370
	$Li_{15}Ge_4$	1384	250
Na	NaGe	369	225
K	KGe	369	Not known

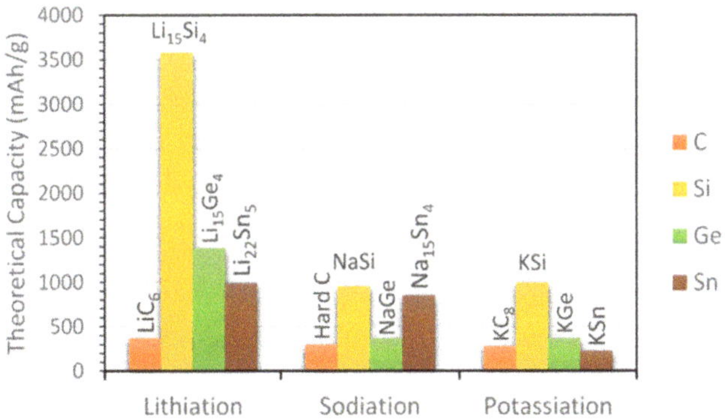

Figure 3.1. Theoretical lithiation, sodiation and potassiation capacities of C, Si, Ge, and Sn. Source: [11–13]. Reproduced with permission of Wiley-VCH Verlag and American Chemical Society.

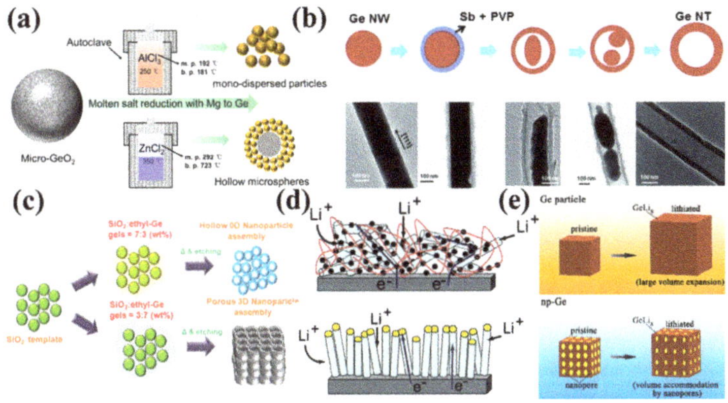

Figure 3.2. (a) Schematic illustration of synthesis of Ge particles and hollow Ge microspheres by metallothermal reduction. Source: [15]. Reproduced with permission of American Chemical Society. (b) Illustration of the synthetic route for Ge nanotubes utilizing the Kirkendall effect and corresponding TEM images for each synthetic step. Source: [16]. Reproduced with permission of John Wiley and Sons. (c) Preparation of 0D hollow and 3D porous nanoparticles by sacrificial template method. Source: [17]. Reproduced with permission of John Wiley and Sons. (d) Rational design of Ge nanowires via vapor–liquid–solid growth directly on the current collector. Source: [18]. Reproduced with permission of American Chemical Society. (e) Illustration of the lithiation for Ge particle and the nanoporous Ge structure. Source: [19]. Reproduced with permission of Elsevier B.V. and Science Press.

The Qian [15] group developed a strategy of low-temperature molten salt system for metallothermal reduction to produce nanostructured Ge, and bulk GeO_2 could be reduced to nanostructered Ge in metallic Mg in molten $ZnCl_2$ and $AlCl_3$, as shown in figure 3.2(a). The morphology characteristics are in the appearance of particles and hollow-spheres, respectively. The reaction temperature and the physical properties of the molten salts played key factors in determining the morphology and structure of the samples. The hollow Ge microspheres deliver an enhanced electrochemical Li-storage performance with a specific capacity of 1217 mAh g^{-1} after 400 cycles at 0.8 C. The great cycling stability benefits from that the hollow structure could easily release the strain stress caused by volume change during lithiation/delithiation processes, and at the same time, the high specific surface and active sites could promote ion transformation, enhancing the rate capability. An efficient synthetic method for fabricating ultralong Ge nanotubes was presented using the Kirkendall effect (figure 3.2(b)) [16]. The hollow structure inside amorphous Ge nanotubes provides buffer space for volume expansion and allows for homogenous volume change during Li alloying and dealloying reactions. It could remain at 70% capacity retention (corresponding to 716 mAh g^{-1}) for both charge and discharge between 3 and 4.3 V at 40 C (or 40 A g^{-1}). The superior performance is due to the tubular morphology with uniform wall thickness created from the unique synthetic strategy by Kirkendall effect. Cho *et al* [17] firstly reported a method of controlling the amount of template and precursors to synthesize 0D and 3D Ge porous nanoparticles, as shown in figure 3.2(c). Initial discharge capacities of the 0D and 3D sample anodes were 1428 and 1380 mAh g^{-1}, and the cycle-life property could maintain 100 cycles at a rate of 1 C with capacity retention ratio of 85% and 95%, respectively. After cycling, there is only a slight increase in wall thickness of 3D Ge with an almost complete preservation of the 3D pore structure, while the 0D Ge shows a fully destruction of the hollow morphology. The long-range ordering of the pore and regularity of the pore wall thickness largely determines the electrochemical reaction process.

Different from traditional anodes for LIBs, Cui *et al* [18] provided a fabricated strategy of Ge nanowires grown on current collector directly as anode, as shown in figure 3.2(d). There is a good interface contact between current collector and each nanowire, which could avoid the risk of electrical contact lost caused by large volume changes during reactions. Meanwhile, 1D Ge nanowires could boost the transport of Li^+ from the electrolyte and electrons from the current collector, and the small diameter allows facile volume changes without fracture. Feng's group [19] prepared nanoporous Ge by chemical dealloying. The lithiation for Ge is shown in figure 3.2(e), and the large volume expansion occurred in Ge particle anode account of Li insertion, which may pulverize the electrical contact between Ge and the current collector. Moreover, elastic strain with Li insertion could worsen the kinetics of Li compound formation. In contrast, the pores in the nanoporous Ge can accommodate the volume change and release the strain. Furthermore, the large surface area could promote the Li-ion/electron exchange efficiently, which improves the high-rate capability and cycle life.

Expect for these, various structures of Ge were created to improve Li^+ kinetics and provide space to accommodate the volume expansion after lithiation. Nanowires [20, 21], nanotubes [22, 23], nanospheres [15], nanoarrays [24], and nanoparticles [25, 26] have attracted lots of attention for their great strain relaxation ability, short diffusion paths for Li^+, stable interfacial contact. However, the fabrication of pure Ge nanostructures is not sufficient to meet future battery performance needs, and carbon matrices were normally introduced to construct Ge–carbon composites as volume buffers to modify their electrochemical performance [27].

3.1.1.2 Ge–C composites

Guo *et al* [28] reported a new strategy by encapsulating Ge nanoparticles in carbon nanoboxes, as shown in figure 3.3(a). This attenuated the volume effect and improved electronic conductivity via providing extra void space for the expansion of Ge nanoparticles after Li insertion, and a large reactive area and reduced distance for the Li^+ diffusion were also introduced. Wan *et al* [29] designed a Ge@C nanocomposites with a carbon nanocoating and RGO (reduced graphene oxide) network, successfully improving both the cyclability and rate capacity (figure 3.3(b)). The double protection of C shell and elastic RGO 3D network prevented direct contact between the particles and the electrolyte, relieving volume changes of Ge anode. Figure 3.3(c) shows the building process of a 3D interconnected porous nitrogen-doped graphene foam (NGF) with encapsulated Ge quantum dot@nitrogen-doped graphene yolk–shell nanoarchitecture (Ge-QD@NG/NGF). The yolk–shell nanoarchitecture composed by N-doped graphene outer shell and Ge quantum dots afforded internal void space for the minimal volume expansion during lithiation, minimizing the pulverization, exfoliation and aggregation of Ge anode [30]. Guo *et al* [31] fabricated a hierarchical micro-nanostructured

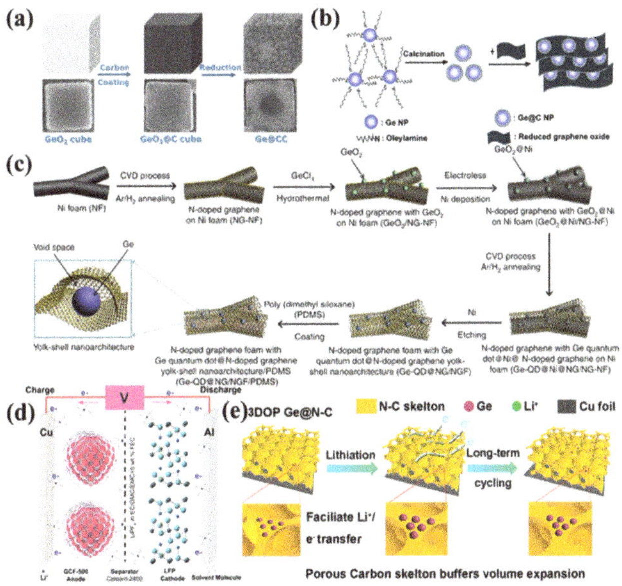

Figure 3.3. (a) Formation process for Ge@carbon cubes and the corresponding SEM images. Source: [28]. Reproduced with permission of John Wiley and Sons. (b) Synthesis route for the Ge@C core–shell nanoparticles and graphene networks nanocomposite. Source: [29]. Reproduced with permission of American Chemical Society. (c) Schematic illustration of 3D nitrogen-doped graphene foam with encapsulated Ge/nitrogen-doped graphene yolk–shell nanoarchitecture. Source: [30]. Reproduced with permission of Nature Publishing Group. (d) Sketch of full-cell configuration with Ge–C framework anodes. Source: [31]. Reproduced with permission of John Wiley and Sons. (e) Schematics of the lithiation and subsequent delithiation process of the Ge quantum dots in 3D ordered porous carbon frameworks. Source: [32]. Reproduced with permission of John Wiley and Sons.

Ge–C framework (GCF) designed by facile and scalable structural engineering strategy (figure 3.3(d)). Hierarchical Ge–C microstructures ensure high vibrational density and electrical conductivity for the low total cost of the composite anode to meet the practical requirements. Figure 3.3(e) shows a binder-free anode comprising Ge quantum dots encapsulated in a 3D ordered porous (3DOP) N@C framework. Well dispersed Ge quantum dots with the particle size of 2–5 nm and high mass loading of as high as 82.9% facilitated the active Li^+ reacting completely with Ge at high lithiation reaction rates, and simultaneously the 3DOP carbon shell can effectively release the volume expansion during cycling [32].

3.1.1.3 Ge–X composites

In addition to the carbon supports, many other materials could be used in Ge-based composites to improve their electrochemical properties as anode materials [33]. It is demonstrated to be a useful way to suppress the volume effect of Ge by introducing other inactive ingredients and creating the composites composed of a fine dispersed inactive matrix of Ge. Typically, the inactive components could act as buffers and relieve the volume changes of Ge during cycling, thus preventing the battery anode from being crushed in LIBs. The matrix needs to promote the transfer of Li^+ and electrons, enhancing the electrochemical performance of Ge anodes. Besides, the inactive matrices also need to display a high mechanical strength to accommodate the volume change of Ge anodes.

Kim *et al* reported [34] a one-pot way to synthesize a spherical structure composed of different ratios of reduced Ge as active component and amorphous TiO_2 as inactive matrix (figure 3.4(a)). Compared with carbon-based material, amorphous TiO_2 matrix could alleviate the volume expansion of Ge anodes without reacting with Li^+ in the lithiation and delithiation processes. The relaxed volume change in Ge/TiO_2 nanocomposites can prevent the pulverization/delamination and inhibit the formation of SEI layer. Therefore, compared to commercial Ge anode, Ge/TiO_2 anode exhibits excellent electrochemical properties, including higher rate performance with larger coulomb efficiency and longer cycle life. Fang *et al* [35] also fabricated Ge/TiO_2 composite by a facile two-step way, and it delivers an great cycling stability of 600 cycles at 5 A g^{-1}, with a reversible specific capacity of 700.3 mAh g^{-1} remained.

Ge/MoS_2 nanocomposite was prepared by Hsieh *et al* [36] via mixing the Ge nanoparticles with the MoS_2 nanosheets in a solution, followed by annealing at 350 °C. Because of the low binding force between the S–Mo–S layers, Li^+ could be easily deintercalated and intercalated into the structure of MoS_2 layers. The addition of Ge nanoparticles to this compound can effectively prevent the restacking of the MoS_2 nanosheets and improve the structural stability and electrical conductivity of the anode. In the aspect of Ge, the problem of large volume variation in the production process was effectively solved through the MoS_2 framework, into which Li^+ was allowed to insert, as depicted in figure 3.4(b). At 0.1 A g^{-1}, the initial discharge and charge capacities of Ge/MoS_2 anode are 1451 and 1147 mA h g^{-1}, respectively. At 0.2 A g^{-1}, the Ge/MoS_2 anode delivers a large reversible capacity of 1362 mA h g^{-1} after 50 cycles.

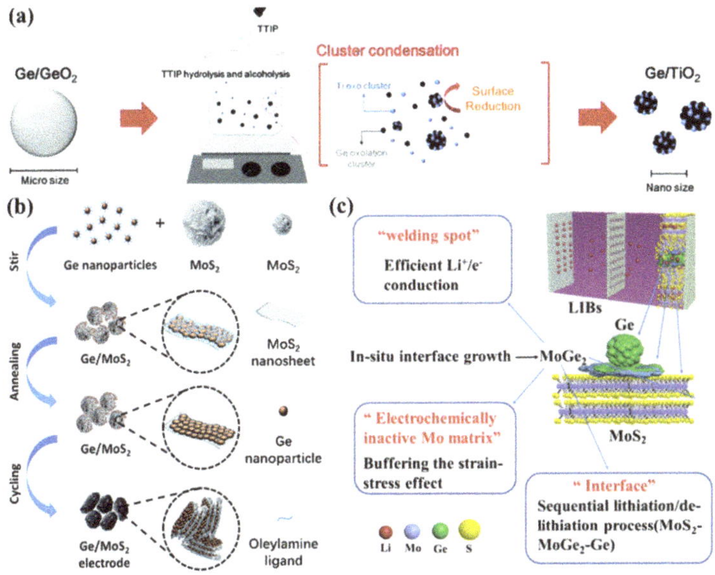

Figure 3.4. (a) Schematic demonstration of the synthesis process of Ge/TiO_2 composite. Source: [34]. Reproduced with permission of Royal Society of Chemistry. (b) Illustration of a facile process the Ge/MoS_2 composite and its structure evolution during cycling. Source: [36]. Reproduced with permission of Royal Society of Chemistry. (c) Schematic illustration and the multi-fold merits of the as-proposed *in situ* interface growth strategy of MoS_2–$MoGe_2$–Ge. Source: [37]. Reproduced with permission of Elsevier B.V. and Science Press.

Li *et al* reported [37] an *in situ* interfacial growth method, in which a strong and conductive $MoGe_2$ alloy was introduced between flexible MoS_2 nanosheets and electrochemically active Ge nanoparticles to regulate their Li storage kinetics. As shown in figure 3.4(c), hierarchical structure was assembled by MoS_2 nanosheets, $MoGe_2$ interfaces and ultrafine Ge nanoparticles at the micro scale. The Ge atoms at the interface of $MoGe_2$ belong to both the alloy phase of $MoGe_2$ and the metallic phase of Ge, which connected the alloy phase of $MoGe_2$ and the metallic phase of Ge. The Mo atoms at the $MoGe_2$ interface shared with both the original MoS_2 phase and the $MoGe_2$ alloy phase. At the molecular level, the heterogeneous interface between high conductive scaffolds and highly capacity Ge can effectively remission the stress–strain effects in charge and discharge processes. Besides, highly Li^+/ e^- electrically conducting $MoGe_2$ interface, acting as a 'solder joint', connected the MoS_2 nanosheets and the Ge nanoparticles, which thus ensures the rapid charge transfer and boost the electrochemical reactions. The Ge@$MoGe_2$@MoS_2 anodes exhibit a superior rate capability of 1268.9, 1136.7, 1082.6, 1031.4, 970.5 and 866.7 mA h g^{-1} at current density of 0.1, 0.2, 0.5, 1, 2 and 5 A g^{-1}, respectively. The fancy designed Ge@$MoGe_2$@MoS_2 anode with a promising tap density of 1.23 g cm^{-3} was proved to show superior battery performance, including the specific capacities of 866.7 mAh g^{-1} at 5.0 A g^{-1} and reversible capacity of 838.5 mAh g^{-1} after 400 cycles. Additionally, rational design of other Ge–X composites was also conducted in recent years. Choe *et al* [38] fabricated a composite structure of

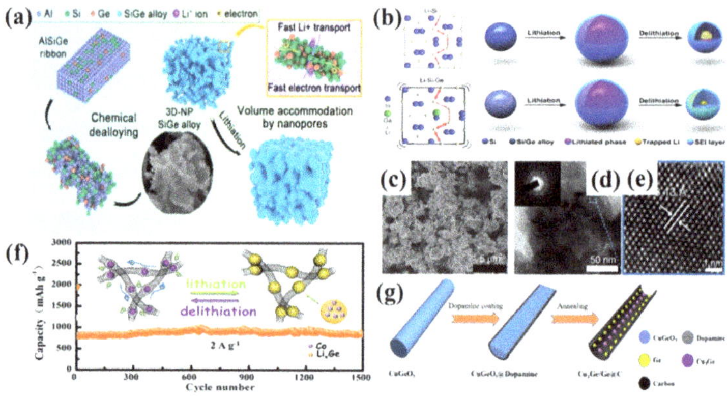

Figure 3.5. (a) Preparation and lithiation/delithiation processes of 3D-NP SiGe. Source: [41]. Reproduced with permission of American Chemical Society. (b) Li diffusion and lithiation/delithiation processes in Li–Si and Li–Si–Ge anodes. Source: [42]. Reproduced with permission of American Chemical Society. (c) SEM, (d) TEM and (e) HRTEM images of $Si_{15}Ge$ nanoparticles (inset: selected area electron diffraction). (f) Cycle performance and proposed lithiation/delithiation process of the Co_5Ge_3/N-CNT composite anodes. Source: [49]. Reproduced with permission of American Chemical Society. (g) Fabrication process of peapod-like Cu_3Ge/Ge@C. Source: [44]. Reproduced with permission of John Wiley and Sons.

Ge/Ti_4O_7 as the anode for LIBs, which shown a low charge transfer resistance after long-term repeated circulation. Kim *et al* [39] prepared ZnO nanoparticles coated Ge nanoparticles (Ge@ZnO NPs), and it could deliver more than 500 stable cycles.

3.1.1.4 Ge-based alloys

Since 2000, the alloying strategy for Ge-based materials has also attracted much attention. It was found that compared with pure Ge anode, Ge-based alloys with different architecture could cope with the volume variations during the lithiation/delithiation processes, showing excellent electrochemical performance, such as Ge–Si [40–42], Ge–Cu [43, 44], Ge–Sn [45] and others [46, 47]. For the Ge–Si alloy, Yang *et al* [41] used a facile, low-cost, and practical dealloying method to fabricate three-dimensional nanoporous (3D-NP) materials for LIBs, the ternary AlSiGe ribbon and Al acted as the precursor and sacrificial element, respectively, as shown in figure 3.5(a). The morphology and porosity of the Ge–Si alloy can be regulated according to the content of Al in the precursor. It was found that 3D-NP SiGe emerges uniformly coral-like structure with continuous ligaments and hierarchical micropores and mesopores, which leads to a high reversible capacity of 1158 mAh g^{-1} after 150 cycles at 1000 mA g^{-1} with excellent rate capability. As depicted in figure 3.5(b), Zhu *et al* [48] synthesized the Si–Ge alloy particles with various atomic ratios using a convenient ball milling method, and corresponding morphology and reaction mechanism are listed in figures 3.5(b)–(e). They demonstrated that isovalent isomorphism can be used to remarkably reduce the energy barriers of Li^+ diffusion and therefore reduce Li^+ trapping and increase initial columbic efficiency. In addition, adding inactive metal into Ge can improve electrical conductivity and reduce volume expansion, contributing to superior electrochemical behavior.

Wu *et al* [49] used ZIF-67 and GeO_2 as precursor to synthesize a nitrogen-doped CNT encapsulated Co_5Ge_3 alloy (Co_5Ge_3/N-CNT) composite (figure 3.5(f)). Compared with Ge metal, Co_5Ge_3 alloy shows lower interaction energy with Li and higher electrical conductivity, which is more conducive to Li^+ diffusion. It is demonstrated that Co_5Ge_3/N-CNT anode exhibits a high reversible capacity of 853.7 mA h g^{-1} and an attractive rate performance of up to 10 A g^{-1} after 1500 cycles of 2 A g^{-1}. As shown in figure 3.5(g), Wang *et al* [44] synthesized a pod-like Cu_3Ge/Ge@C composite by annealing $CuGeO_3$@dopamine, and the introduction of carbon and inactive Cu phase enhanced the electrical conductivity of Ge, acting as a buffer to reduce the volume change in the lithiation/delithiation processes. It also exhibits long cycle life, with a discharge specific capacity as high as 934 mAh g^{-1} after 500 cycles. Chang *et al* [50] prepared stacked Ge/Cu nanowires (NWs) laminates, which makes the nanowire layer structure strong and compact, and the conductive adhesive prevents the formation of Cu_3Ge, and the stacked Ge/Cu nanowire laminate anode with surface area of 14.8 mg cm^{-2} delivers a reversible capacity of 16 mAh per unit area with Ge utilization of approximately 80%.

3.1.1.5 Binary and ternary Ge compounds

Binary/ternary Ge compounds also hold great promise as anodes for LIBs. In order to improve the electrochemical performance and cope with the volume change during the reaction, the research on GeO_x focuses on the construction of different nanostructures, and the composite of GeO_x with other matrices to maximize the advantages of Ge. Seng *et al* used a simple hydrothermal method to synthesize GeO_2 nanoparticles (GeO_2-nano), and GeO_2-nano was annealed in an acetylene catalyst/hydrogen atmosphere to form a uniform carbon coating on the particles at 520 °C, resulting in GeO_2/C. When the temperature increased to 650 °C, part of GeO_2 reacted with carbon to obtain GeO_2/Ge/C. This kind of nano-sized particles can effectively deal with volume changes and enable close contact of Ge and LiO_2, which is more conducive to the electrochemical reaction. In addition, the introduction of carbon can not only increase the reversibility and electrical conductivity, but also enhance the reaction kinetics of Li insertion and deinsertion. It can also be used as a buffer matrix to relieve volume changes during charge and discharge processes. Interestingly, the element Ge in GeO_2/Ge/C acted as a catalyst for facilitating the decomposition of Li_2O, which is favorable for improving rate performance (figure 3.6(a)) [51]. As illustrated in figure 3.6(b)–(d), Zhang *et al* [52] successfully developed the scalable production of 2D GeO_x@MXene composite through simple wet chemical methods at room temperature. It features high electrical/ionic conductivity and better volume adjustment, which results in excellent rate capability up to 20 C. Other binary/ternary Ge compounds were also attempted as anode candidates for LIBs, including $CuGeO_3$, [53] $Ca_2Sb_2Te_5$, [54] $Ca_2Ge_7O_{16}$, [55] Co_2GeO_4, [56] Zn_2GeO_4 [57] and GeP_5, [58, 59] Ge_3N_4, [60] GeS_2 [61], GeSe, [62] GeTe [63], etc. Li *et al* [59] synthesized the layered GeP through a simple ball milling process and low-temperature annealing, inspired by the small difference in electronegativity between Ge and P. It exhibits better LIB performance than those of Ge anodes, which is due to the better Li-storage kinetics and the synergistic effect of Ge and P.

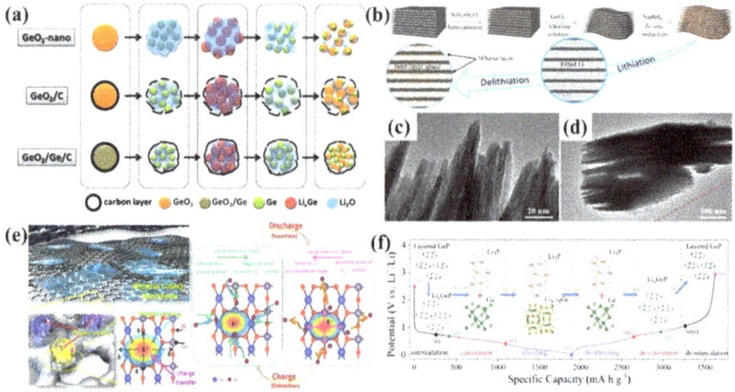

Figure 3.6. (a) Proposed charge/discharge process of $GeO_2/Ge/C$ anode. Source: [51]. Reproduced with permission of American Chemical Society. (b) Schematic of the synthesis process of GeO_x@MXene. TEM images of (c) MXene and (d) GeO_x@MXene. Source: [52]. Reproduced with permission of American Chemical Society. (e) Reaction mechanism of $CuGeO_3$/graphene anode. Source: [53]. Reproduced with permission of American Chemical Society. (f) Reaction route of monoclinic phase GeP anode. Source: [59]. Reproduced with permission of Elsevier B.V. and Science Press.

In addition, with the low formation energy, it has excellent self-repair ability in rebuilding the original layered structure during cycling, and this process was proposed and listed in figure 3.6(f). Liu *et al* [53] used the traditional liquid phase exfoliation method to fabricate 2D $CuGeO_3$/graphene materials as a high-performance anode material for LIBs. They found that oxygen vacancies formed when the thickness dimension was reduced to the atomic level during the preparation process. The introduction of vacancies realized the nearby charge distribution imbalance and generated a local electric field, which accelerated the ion and electron migration rate (figure 3.6(e)). The synthesized unique structure could make full use of the advantages of Ge-based compounds to alleviate the problem of volume expansion, obtaining a long cycle life of 500 cycles and a reversible specific capacity of 693 mAh g^{-1} at 1000 mA g^{-1}.

3.1.2 Ge anode for Na-ion batteries

3.1.2.1 Ge nanostructures

According to previous reports, Ge-based materials have been extensively studied as anodes for LIBs, as they feature high Li storage capacity and high Li^+ diffusivity [64–66]. However, the Li still has some problems, including limited resource and uneven distribution, which lead to a high price and in turn limits its application in large-scale [67, 68]. Due to the similar electrochemical properties to Li, large reserves and low prices, SIBs have attracted more and more attention and been considered as the promising next-generation energy storage techniques [69, 70]. Generally, materials with good performance in LIBs can often also be attempted in SIBs [71]. Therefore, Ge-based materials were also intensively studied as anodes for SIBs, on the alloying mechanism in the form of NaGe with a specific capacity of 369 mAh g^{-1} [72–74]. However, the cycling performance and rate property of Ge anode for SIBs remain limited. It is noted that Na exhibits larger ionic radius than Li (0.102 nm for Na^+ vs. 0.076 nm for Li^+), which could lead to sluggish reaction kinetics and poor cycling stability. In addition, the anode with alloying mechanism also tent to induce large volume change during cycling, and this would also cause collapsion and pulverization of active materials, posing a huge challenge to the practical application of Ge in SIBs. In recent years, great efforts have been engaged in dealing with these problems through designing specific structure with various morphology and compounding other substances.

Na^+ diffusion in crystalline Ge are interstitial diffusers. As mentioned earlier, the radius of Na^+ is much larger than that of Li^+, which results in a much higher activation energy for the ions hoping between two interstitial sites in the lattice (1.5 eV for Na vs. 0.51 eV for Li). Sluggish solid-state diffusion of Na^+ is known as a significant fraction of the total cell resistance. From the previous reports, the grain boundary diffusion and surface diffusion are both much faster than the bulk diffusion in solid for SIBs. Both nanocolumnar and dense Ge thin film were investigated by Mullins *et al* [75] as potential SIB anodes, synthesized via an evaporative deposition method. Ge in the form of nanocolumnar film exhibits much better electrochemical property than the dense film counterpart, with a specific capacity of 430 mAh g^{-1}, retaining 88% of their initial capacity after 100 cycles at C/5. It was stably charged and discharged at up to 27 C (10 A g^{-1}), while the dense film began to fade after 15 cycles. Electrochemical impedance spectroscopy analysis was conducted, and the results demonstrate that the Na^+ diffusion in the near-surface layer is significantly faster than that in the bulk diffusion. As a result, it is critical to utilize nanostructured Ge to increase the surface and grain boundary Na^+ diffusion in Ge, and nanoscale material design is thus critical for rapid, reversible, and high-rate sodiation of Ge.

Kim *et al* [76] synthesized Ge nanoparticles with different diameters by a laser pyrolysis method, and the average diameter of these nanoparticles could be easily tuned by changing sensitizer gas flow rates in the preparation the process. It was assumed that the small physical size itself alongside the improved mechanical properties of smaller nanoparticles higher the resistance against anode failure during cycling, and the battery performance was thus improved as the size of Ge powder decreased consistently. Gavrilov's group prepared Ge nanowires by electrodeposition from aqueous solution at different temperatures, which were used as anodes for SIBs [77]. They display discharge capacities of 100 and 370 mAh g^{-1} at 2 and 0.4 A g^{-1}, respectively, and 90% of the initial revisable capacity after 50 cycles could also be obtained [77]. Qian's group synthesized amorphous Ge by acid-etching Mg_2Ge (figures 3.7(a)–(b)) with the surface area of 29.9 m^2 g^{-1} and average pore diameter of 7.6 nm [78]. This amorphous Ge showed superior Na storage property, including reversible specific capacity of 550 mAh g^{-1} at 0.2 C after 50 cycles, favorable rate capability of 273 mAh g^{-1} at 5 C and stable cycling performance of more than 200 cycles at 0.4 C. This study indicates that the mesoporous structure could enhance the Na^+ diffusion rate and relieve the volume change during cycling, evidenced by the anodes of pristine and after 50 cycles in figures 3.7(c) and 3.7(d). It is known that the delithiation could trigger nanoporosity for the anode based on

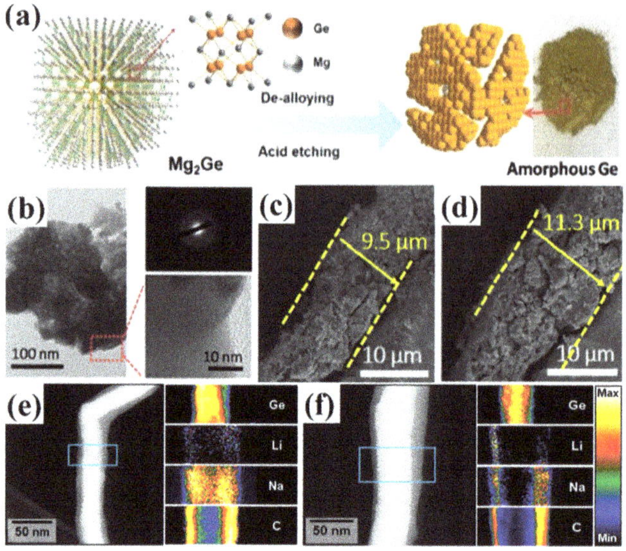

Figure 3.7. (a) schematic representation of the fabrication process of the mesoporous amorphous Ge. (b) TEM image with SAED, the thickness of the anodes of (c) pristine and (d) after 50 cycles. Source: [78]. Reproduced with permission of Springer. (e)–(f) TEM and elemental mapping images of Ge nanowires after initial sodiation and desodiation. Source: [74]. Reproduced with permission of American Chemical Society.

insertion and conversion mechanisms, including Si- and Ge-based films, nanowires, nanoparticles and so on [79, 80]. Mitlin *et al* improved SIB performance of Ge nanowires and thin films by applying single lithiation/delithiation cycle. It was traced that the volume changes before and after the lithiation process is not obvious (figures 3.7(e) and 3.7(f)), mainly due to the activation with inducing amorphization in Ge nanowires, decrease of the barriers for nucleates of NaGe phase and acceleration of the solid-state diffusion. After Li activation, the nanowire and thin film anodes could exhibit 355 and 360 mAh g^{-1} at 1 C and 283 and 310 mAh g^{-1} at 4 C, respectively [78].

3.1.2.2 Ge–C composites

As we know, carbon materials show the merits of holding a wide range of sources, good electrical conductivity, easy structure design and easy access to porous materials with various morphologies. If the Ge material is nanostructure and composited with the void carbon material, in theory, its electrochemical performance would be even much better. Ge nanoparticles were encapsulated in interconnected hollow carbon boxes (Ge/C hybrid) by Yin *et al* [81] using $NH_4H_3Ge_2O_6$ as the precursor, which were employed as anodes for SIBs (figures 3.8(a)–(c)). It could remain a specific capacity of 346 mAh g^{-1} after 500 cycles at 100 mA g^{-1}, close to the theoretical capacity of 369 mAh g^{-1}. These hollow carbon boxes can effectively alleviate the volume expansion of the Ge material during Na^+ insertion/extraction processes, thereby greatly improving the cycle stability and reducing the structure pulverization and collapse of the active materials. Duan *et al* [82] designed an

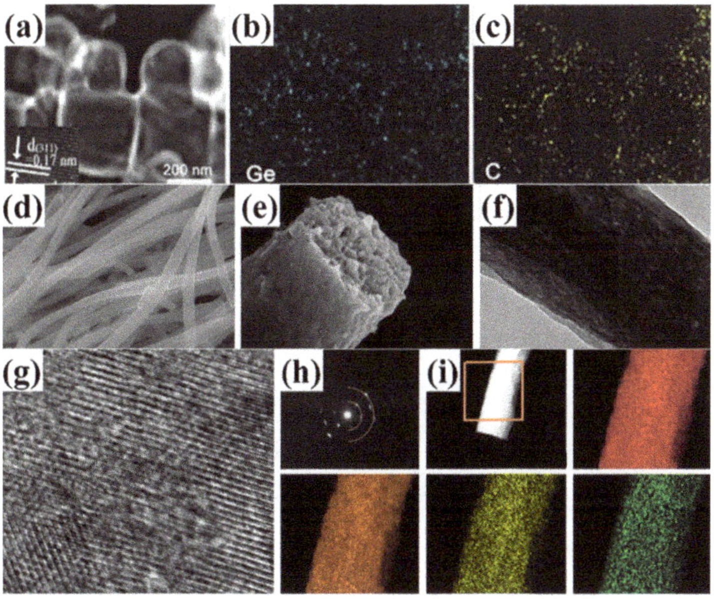

Figure 3.8. (a) TEM and (b, c) EDX elemental mapping images of Ge/C hybrid. Source: [81]. Reproduced with permission of John Wiley and Sons. (d) SEM image of NMCFs, (e) cross-sectional SEM, (f) Low-magnification TEM (g) HRTEM, (h) corresponding SAED and (i) EDS element mapping images of Ge-NMCFs. Source: [82]. Reproduced with permission of IOP Publishing.

ultrafine Ge nanoparticles embedded in hierarchical N-doped multichannel carbon fibers (Ge-NMCFs) through electrospinning, and their morphologies and NMCF architecture are depicted in figures 3.8(e)–(i) and figure 3.8(d), respectively. These carbon nanofibers can not only relax the volume expansion problem of Ge materials, but also provide a large number of channels for the Na^+ transmission and accelerate the battery reaction kinetics. It exhibits an initial specific capacity of 303 mAh g^{-1} and remained 160 mAh g^{-1} after 700 cycles in SIBs.

3.1.2.3 Ge–X composites

Wang *et al* [83] prepared Ge@graphene@TiO_2 core–shell nanofibers (Ge@G@TiO_2 NFs) through the electrospinning and atomic layer deposition method, and it was used as the anode materials for SIBs. Its average diameter is about 100 nm and the surface is not smooth, on which a large number of nanoparticles were loaded, as evidenced in figures 3.9(a) and (b). EDS elemental mapping images show that Ge, Ti and O elements are uniformly distributed on the nanofibers. Figure 3.9(c) shows a highly ordered nanocrystalline Ge with a lattice fringe of 0.32 nm, corresponding to the (111) plane. Ge@G@TiO_2 NFs anode delivers a high specific capacity over 190 mAh g^{-1} after 100 cycles with a high columbic efficiency of around 99% and excellent rate performance, compared with those of Ge@G and bare G anodes. It is demonstrated in figure 3.9(d) that graphene acted as a conductive network to offer more active sites and effective channels for electron and ion transportation, and the coated ultrathin TiO_2 could stabilize the SEI layer and maintained the integrity of

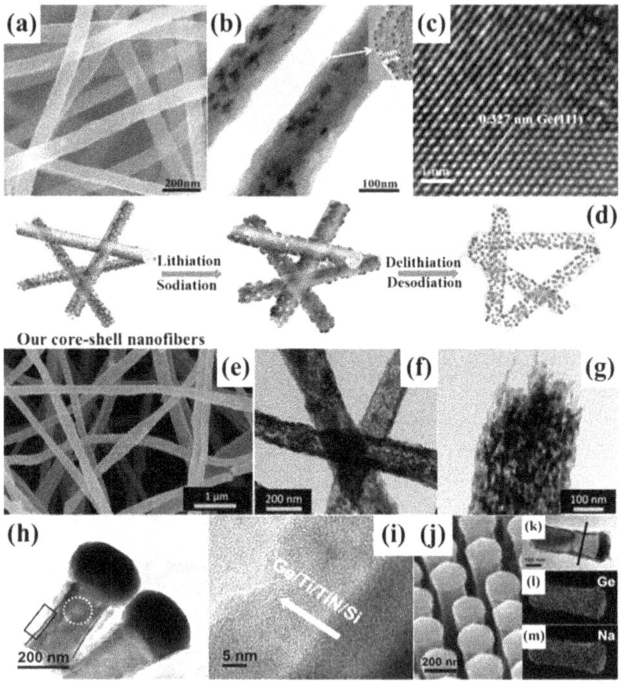

Figure 3.9. (a) SEM, (b)TEM and (c) HRTEM images of Ge@G@TiO_2 nanofibers. (d) Volume change in the core–shell NFs before and after cycle. Source: [83]. Reproduced with permission of American Chemical Society. (e) SEM and (f)–(g) TEM images of Fe_2O_3–Ge nanofibers. Source: [84]. Reproduced with permission of MDPI. (h) TEM and (i) HRTEM images of the Si/TiN/Ti/Ge composite nanorods. (j) SEM and (k)–(m) of EDS elemental mapping images taken from the composite anodes after 60 cycles. Source: [85]. Reproduced with permission of John Wiley and Sons.

the composite anode. Recently, Fe_2O_3–Ge nanofibers composite with similar morphology was prepared by Petrovičovà *et al* [84], figure 3.9(e) shows the diameter of smooth nanofibers ranges from 65–1180 nm. Compared with the pure Fe_2O_3 nanofibers, Fe_2O_3–Ge nanofibers display finer and longer nanostructures, and they grow mainly along the longitudinal axis of the fibers, as pictured in figures 3.9(f) and 3.9(g). The anodes show good specific capacity of 320 mAh g^{-1} at 50 mA g^{-1} and excellent rate capability of 140 mAh g^{-1} at 2 A g^{-1} for SIBs. This is mainly due to the synergistic combination of the electronic transport properties of the composite material, the nanostructured morphologies and the pseudo-capacitive nature of the charge storage mechanism. Yue and co-workers [85] fabricated 3D Si/Ge nanorods buffered by TiN/Ti interlayer as anode for SIBs. Figure 3.9(h) displays the match-like Si/TiN/Ti/Ge composite nanorods. It is clear that the Si core is in close contact with the shell of TiN/Ti/Ge, and the thickness of the whole core–shell structure is about 30 nm (figure 3.9(i)). The deposited Ge layer is amorphous, owing to the fact that no lattice fringes were observed on it. After ultrasonic cleaning process in alcohol, the Ge layer was successfully removed off the composite anode after 60 cycles, and only thin and close contact Ge film around Si nanorods was left, as shown in figure 3.9(j). Corresponding EDS element mappings (figures 3.9(k)–(m))

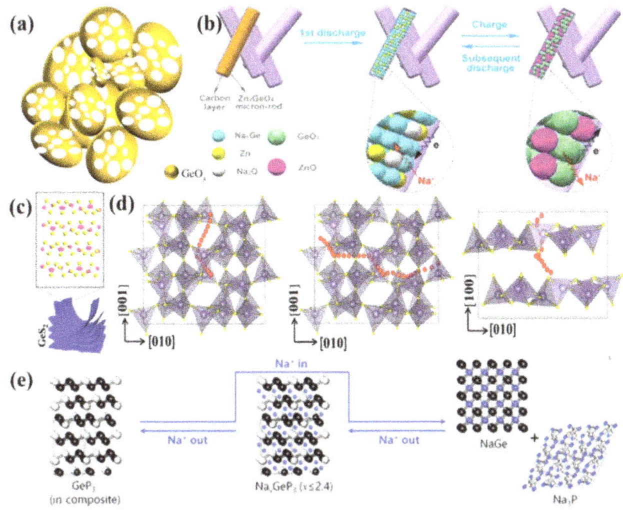

Figure 3.10. (a) Sketch of the GeO_x. Source: [88]. Reproduced with permission of The Royal Society. (b) Schematic illustration of the discharge and charge processes for Zn_2GeO_4@C micron-rods anode. Source: [105]. Reproduced with permission of Elsevier B.V. and Science Press. (c) Schematic demonstration of the ultrathin GeS_2 nanosheets with (d) diffusion pathways viewed along different directions. Source: [93]. Reproduced with permission of American Chemical Society. (e) Schematic diagram of the electrochemically-driven crystallographic phase-change mechanism of GeP_3 during Na insertion/extraction. Source: [106]. Reproduced with permission of Elsevier B.V. and Science Press.

revealed Na and Ge elements are uniformly distributed on the surface of the composites NRs after cycling, indicating that Na^+ were preferentially inserted into Ge active materials. The three-dimensional amorphous Ge material and the effective TiN/Ti electron-conducting layer are effective for the good cycling stability of Si/TiN/Ti/Ge anode.

3.1.2.4 Ge-based alloys

Currently, Ge-based compound materials have also received extensive research interest as suitable anodes for SIBs, such as Ge-based oxides, [86–90] chalcogenides, [91–94] phosphides, [58, 95–99] carbides, [100] and corresponding composites with carbon materials [101–104]. In order to solve the volume expansion problem, which usually lead to unfavorable electrochemical performance after sodium storage, researchers found that it is an effective way to prepare amorphous oxide anodes and spinel structure oxide to serve as SIBs anodes. As shown in figure 3.10(a), Shen's group [88] synthesized amorphous GeO_x with a hierarchical mesoporous structure at room temperature through a chemical reduction route. Thanks to the amorphous hierarchical porous structure, the volume changes caused by the Na storage was greatly alleviated, thereby it can deliver the excellent specific capacities of 200 mAh g^{-1} even at 3.0 Ag^{-1}. In addition, spinel oxides such as Zn_2GeO_4 [89] and $NaGe_{3/2}Mn_{1/2}O_4$ [90] can also achieve highly reversible Na insertion/deletion due to their large-sized tunnel, thus exhibiting extremely low volume changes and high cycle stability. However, Ge-based oxides still show some fatal defects, and especially the lower electrical conductivity would seriously hinder the charge transfer. Compositing carbon materials to modify the surface of oxides was thus widely applied to improve their electrical conductivity [103, 104]. For example, Li's team [105] synthesized carbon-coated Zn_2GeO_4 micron rods (Zn_2GeO_4@C) by a simple hydrothermal method. Due to the improved electrical conductivity, it delivers a high reversible capacity of 317 mAh g^{-1} with good capacity retention of 88%. Figure 3.10(b) shows a schematic diagram of the corresponding electrochemical energy storage process. During the initial discharge, the crystal structure of Zn_2GeO_4 was destroyed, and a further alloying reaction ($Ge + xNa^+ + xe^- \rightarrow Na_xGe$) occurred between Na and Ge. In the subsequent charge and discharge processes, only the reduction of oxides and the alloying reactions between Na and Ge ($GeO_2 + 4Na^+ + 4e^- \rightarrow Ge + 2Na_2O$ and $ZnO + 2Na^+ + 2e^- \rightarrow Zn + Na_2O$) were experienced. In these processes, the carbon shell would accelerate the transportation of electrons and speed up the reaction kinetics. In fact, in addition to the modification of oxide materials, the preparation of transition metal chalcogenides is also a good choice. Compared with oxides, transition metal chalcogenides display higher electrical conductivity and ability to advance redox kinetics. Li *et al* [93] used a topology transformation route to prepare ultrathin GeS_2 nanosheets (figure 3.10(c)) as SIB anodes, which can provide a high capacity of 515 mAh g^{-1} after being cycled for 2000 times at 10 A g^{-1}. DFT calculation results in figure 3.10(d) clearly show that Na^+ preferably diffused through the [001] and [100] directions, which exhibited the lowest energy barriers of 0.36 and 0.51 V in these two directions, respectively. While in the [010] direction, it shows a very high energy barrier of 1.37 V, along which the Na migration was almost impossible. Moreover, GeS nanosheets [91] and GeSe nanowires [94] have also been proven to be excellent anodes for SIBs. Different from the relatively low electrical conductivity and ion migration speed of Ge-based oxides and S/Se compounds, the Ge–P intermetallic compounds exhibit a layered graphene-like structure, high electrical conductivity and superior carrier mobility, which are considered to be promising materials for SIBs. Nam Zhan *et al* [106] used x-ray absorption fine structure (EXAFS), XRD and other advanced techniques to study the electrochemical phase transition mechanism of the GeP_3-based nanocomposite anode during the Na insertion/extraction processes. As shown in figure 3.10(e), in the Na insertion process, the layered GeP_3 was transformed into NaGe and Na_3P. However, in the Na-extraction process, Na_3P and NaGe were transformed into Na_xGeP_3 ($x \leqslant 2.4$), and then restored to the original GeP_3 through topology, which is similar to the research results of GeP by Shen *et al* [107]

3.1.2.5 Binary and ternary Ge compounds

At present, there is limited research on Ge-based alloys as anodes for SIBs. Abel's group [108] obtained Sn–Ge alloys through a vacuum deposition route with film-like (figure 3.11(a)), and alloys of different compositions were used as the anodes delivering higher cycle stability (figure 3.11(b)). It can be seen that although the stability of the alloy greatly increased due to the Ge introduction, and due to the lower columbic capacity of germanium, the columbic capacity of the alloy decreases as the germanium content increases. In addition, Farbod *et al* [109] successfully prepared Sn–Ge–Sb ternary alloy (figure 3.11(c)), and the electrochemical results in

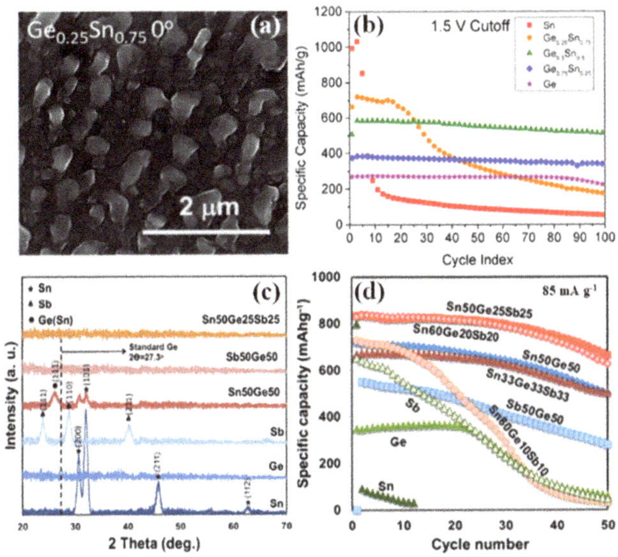

Figure 3.11. (a) SEM images of Sn–Ge alloy films; (b) stability of Sn–Ge alloy film anodes cycled at C/2 between a lower cutoff voltage of 5 mV and an upper cutoff voltage of 1.5 V. Source: [108]. Reproduced with permission of American Chemical Society. (c) Indexed XRD patterns of as-deposited alloyed thin films and (d) their cycling performance at 85 mA g^{-1}. Source: [109]. Reproduced with permission of American Chemical Society.

figure 3.11(d) show that $Sn_{60}Ge_{20}Sb_{20}$ and $Sn_{50}Ge_{25}Sb_{25}$ alloy anodes deliver superior reversible capacity and cycle performance.

3.1.3 Ge anode for K-ion batteries

In order to enhance the electrochemical performance of Ge-based anode materials for PIBs, a lot of effort has been made to fabricate nanostructure, heterostructures and porous materials, which could reserve reaction space and provide high ion and electrical transport conductivity. A effective route for fabricating amorphous Ge anode for potassium-ion batteries was developed by Yi and co-workers [78]. As an anode, this anode could deliver a reversible K-ion storage capacity of 430 $mAh{\cdot}g^{-1}$. The improved performance was mainly attributed to its mesoporous structure. Surface and cross-section structure of the anode after 50 cycles shows that the volume expansion of the anode is not obvious, which indicates the amorphous mesoporous Ge is helpful for restraining volume expansion. Ge@C-CMK composites were prepared by chemical vapor deposition (CVD) method, as shown in figure 3.12(a) [110]. It is worth noting that the dual carbon composites could alleviate the expansion of Ge nanoparticles on the surface of carbon matrix, and the amorphous carbon contributes an extra capacity. Figure 3.12(b) shows the increase of the capacitive capacities with the increase of the scan rates. The reaction mechanism for GeP_5 during potassiation/depotassiation was supposed in figure 3.12(c). At stage 1, GeP_5 decomposed into Ge and P particles, and the P component reacted with K to form K_4P_3. At stage 2, Ge alloyed with K to form KGe.

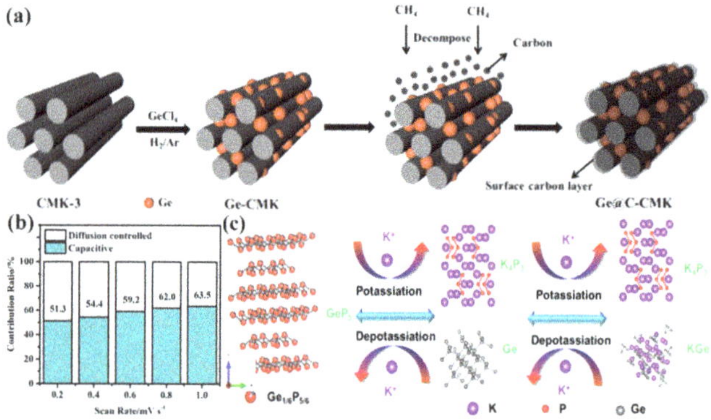

Figure 3.12. (a) Schematic presentation of the synthetic process of Ge@C-CMK composite. (b) The diffusion controlled and capacitive capacities of Ge@C-CMK anodes at different scan rates. Source: [110]. Reproduced with permission of Elsevier B.V. and Science Press. (c) Schematic illustration of potassiation and depotassiation processes of GeP_5 anode. Source: [111]. Reproduced with permission of Elsevier B.V. and Science Press.

Consequently, the K-storage reactions of K–P phases (K_4P_3) and K–Ge (KGe) alloys during the discharge processes can create a mutual buffer mechanism to alleviate the volume changes during cycling, hence aiding in maintaining the anode integrity and enhancing the cycling stability [111]. Yang and co-workers [7] prepared a pure Ge anode with nanoporous structure. This study suggests pure Ge can be a potential anode in KIBs, but the unique porous structural plays a critical role in overcoming the kinetics problems caused by large K^+.

3.2 Challenge of Ge anode for Li, Na, K-ion batteries

Herein, a brief summary on the state-of-the-art advances of Ge-containing materials for clean and next-generation power sources has been presented, including LIBs, SIBs and PIBs. In the past few decades, the electrochemical performance of Ge lithiation and sodiation was intensively investigated by using different preparation methods and various characterization techniques, while the Ge potassiation was highlighted based on the limited available information. Owing to their relatively electrical conductivity, faster ionic diffusivity, large theoretical specific capacity and excellent mechanical strength, Ge-containing anodes hold great potential for the application in these high-energy devices. Notably, a main shortcoming of Ge-containing anodes is that they would show obvious volume expansion during cycling, leading to their pulverization and cracking and poor cycling stability. Moreover, it is evident that the poor electrochemical activities of Ge materials and the unstable SEI layers could also deteriorate the battery performance.

In response, numerous efforts had been devoted to accommodating their volume changes and activating the Ge-containing anodes, and the effective ways could be classified into the following strategies: (i) constructing nano-scaled or/and porous Ge architectures; (ii) incorporating Ge into the conductive carbon matrix; (iii) compositing Ge with other active materials; (iv) fabricating Ge-based alloys; (v) synthesizing binary and ternary Ge-based compounds.

Actually, the electrochemical performance of Ge-containing anodes has been obviously enhanced according to the previous studies, and their structure–activity relationship and reaction mechanism have been investigated in depth. Although great achievements have been achieved, the nanostructure synthesis needs further desirable structural modifications for Ge-based materials with high-performance electrochemical behaviors to realize superior energy and power density and meet the requirements of large-scale applications. To boot the future development, the following researches are highly suggested:

(1) The major challenge for Ge-based materials as commercial anodes at present is to cut down their costs without sacrificing the battery performance. Optimal preparation techniques are needed account of different materials and power source system, including wet chemical reaction, hydrothermal method, vapor–liquid–solid, pyrolysis, chemical vapor deposition and ball milling route. It is proposed that the chemical vapor deposition and wet chemical reaction could be combined to fabricate Ge-based materials considering its simplicity and easy accessibility, which resulting in the effective formation of diverse architectures and structures, and further research into using this industry-standard technique is urgently required.

(2) The SEI layer plays a critical role in improving anode activity and stability, especially the initial columbic efficiency, and its formation during cycling needs to be further controlled. However, this process is usually compositional complex and inhomogeneous, and the generation route on Ge-based anodes and the key influencing factor regulated by the anode morphology and structure remain to be investigated.

(3) Nanostructured materials tend to show high specific surface areas, resulting in the formation of a larger SEI layer with a lower initial columbic efficiency, and the larger boundaries could restrain the long-distance transport of electrons and Li+, lowing the electrical conductivity and ionic diffusivity. Besides, the relatively low tap density of the nanostructured materials could decrease the total anode density and the volumetric capacity, triggering big problem in practical applications. Therefore, developing microsized Ge-based materials is considered as an effective way to deal with the challenge.

(4) Rational design and fabrication of binderless, self-supported, and flexible anodes without the need for other additives are highly required for developing smart and tiny electronic devices. Effective integration of free-standing Ge-based anodes into batteries and how to relax their volume expansion should be focused on in the future research.

(5) Advanced high-resolution characterizations (e.g., atomic-scale observations and structure spectrum measurements) and operando analysis (particularly, *in situ* and/or real-time technologies) are urgently demanded to achieve more insightful perspectives of Ge-based anode structure and mechanisms for the key electrochemical reactions. According to these techniques, the structure–property correlation would be further determined.

(6) Theoretical calculations and simulations are of great importance for obtaining deep insight on the reaction mechanism of Ge-based anodes and the basic principles for the precise design, and construction of advanced Ge-based anodes should be yielded by reducing the gaps between the theoretical calculation data and experimental results.

It is reasonably considered that, with the extensive efforts devoted to the investigation of Ge-based materials, persistent enriched scientific breakthrough would be achieved, and they would play an increasingly active role in the high-performance energy storage devices in the future.

References

[1] Mogus Milankovic A, Sklepic K, Blazanovic H, Mosner P, Vorokhta M and Koudelka L 2013 Influence of germanium oxide addition on the electrical properties of Li_2O–B_2O_3–P_2O_5 glasses *J. Power Sources* **242** 91–8

[2] Danto S, Houizot P, Boussard Pledel C, Zhang X, Smektala F and Lucas J 2006 A family of far-infrared-transmitting glasses in the Ga–Ge–Te system for space applications *Adv. Funct. Mater.* **16** 1847–52

[3] Na H, Kwak D, Kwon Y, Cho H and Kim H 2013 Synthesis of Ge-doped SnO_2 nanowires and their photoluminescence propertiess *J. Ceramic Process. Res.* **14** 391–95

[4] Jung H *et al* 2015 Elucidation of the local and long-range structural changes that occur in germanium anodes in lithium-ion batteries *Chem. Mater.* **27** 1031–41

[5] Kennedy T, Brandon M and Ryan K 2016 Advances in the application of silicon and germanium nanowires for high-performance lithium-ion batteries *Adv. Mater.* **28** 5696–704

[6] Younesi R, Veith G, Johansson P, Edstrom K and Vegge T 2015 Lithium salts for advanced lithium batteries: Li–metal, Li–O_2, and Li–S *Energy Environ. Sci.* **8** 1905–22

[7] Yang Q, Wang Z, Xi W and He G 2019 Tailoring nanoporous structures of Ge anodes for stable potassium-ion batteries *Electrochem. Commun.* **101** 68–72

[8] Lu X, Adkins E, He Y, Zhong L, Luo L, Mao S, Wang C and Korgel B 2016 Germanium as a sodium ion battery material: *in situ* TEM reveals fast sodiation kinetics with high capacity *Chem. Mater.* **28** 1236–42

[9] Qin J and Cao M 2016 Multidimensional germanium-based materials as anodes for lithium-ion batteries *Chem. Asian J.* **11** 1169–82

[10] Xin S, Guo Y and Wan L 2012 Nanocarbon networks for advanced rechargeable lithium batteries *ACC. Chem. Res.* **45** 1759–69

[11] Sultana I, Rahman M, Chen Y and Glushenkov A 2018 Potassium-ion battery anode materials operating through the alloying–dealloying reaction mechanism *Adv. Funct. Mater.* **28** 1703857

[12] Obrovac M and Chevrier V 2014 Alloy negative electrodes for Li-ion batteries *Chem. Rev.* **114** 11444–502

[13] Loaiza L, Monconduit L and Seznec V 2020 Si and Ge-based anode materials for Li-, Na-, and K-ion batteries: a perspective from structure to electrochemical mechanism *Small* **16** 1905260

[14] Liang S, Cheng Y, Zhu J, Xia Y, Müller and Buschbaum P 2020 A chronicle review of nonsilicon (Sn, Sb, Ge)-based lithium/sodium-ion battery alloying anodes *Small Methods* **4** 2000218

[15] Lin N, Li T, Han Y, Zhang Q, Xu T and Qian Y 2018 Mesoporous hollow ge microspheres prepared via molten-salt metallothermic reaction for high-performance Li-storage anode *ACS Appl. Mater. Interfaces* **10** 8399–404
[16] Park M, Cho Y, Kim K, Kim J, Liu M and Cho J 2011 Germanium nanotubes prepared by using the Kirkendall effect as anodes for high-rate lithium batteries *Angew. Chem. Int. Ed.* **50** 9647–50
[17] Park M, Kim K, Kim J and Cho J 2010 Flexible dimensional control of high-capacity Li-ion-battery anodes: from 0D hollow to 3D porous germanium nanoparticle assemblies *Adv. Mater.* **22** 415–18
[18] Candace K, Zhang X and Cui Y 2008 High capacity li ion battery anodes using Ge nanowires *Nano Lett.* **8** 307–9
[19] Gao R, Zhu J, Xiao X, Hu Z, Liu J and Liu X 2015 Facet-dependent electrocatalytic performance of $Co_3O_4$for rechargeable Li–O_2 battery *J. Phys. Chem.* C **119** 4516–23
[20] Collins G, McNamara K, Kilian S, Geaney H and Ryan K 2021 Alloying germanium nanowire anodes dramatically outperform graphite anodes in full-cell chemistries over a wide temperature range *ACS Appl. Energy Mater.* **4** 1793–804
[21] Mullane E, Kennedy T, Geaney H and Ryan K 2014 A rapid, solvent-free protocol for the synthesis of germanium nanowire lithium-ion anodes with a long cycle life and high rate capability *ACS Appl. Mater. Interfaces* **6** 18800–7
[22] Liu X, Hao J, Liu X, Chi C, Li N, Endres F, Zhang Y, Li Y and Zhao J 2015 Preparation of Ge nanotube arrays from an ionic liquid for lithium ion battery anodes with improved cycling stability *Chem. Commun.* **51** 2064–67
[23] Stokes K, Boonen W, Geaney H, Kennedy T, Borsa D and Ryan K 2019 Tunable core–shell nanowire active material for high capacity Li-ion battery anodes comprised of PECVD deposited asi on directly grown Ge nanowires *ACS Appl. Mater. Interfaces* **11** 19372–80
[24] Lee G, Shim H and Kim D 2015 Superior long-life and high-rate Ge nanoarrays anchored on Cu/C nanowire frameworks for Li-ion battery electrodes *Nano Energy* **13** 218–25
[25] Jin S, Li N, Cui H and Wang C 2014 Embedded into graphene Ge nanoparticles highly dispersed on vertically aligned graphene with excellent electrochemical performance for lithium storage *ACS Appl. Mater. Interfaces* **6** 19397–404
[26] Staub J, Dzara M, Crompton K, Ganter M and Landi B 2020 Balancing irreversible capacity in germanium nanoparticle: HE5050 lithium ion batteries for improved energy densities exceeding 250 Wh/kg *Electrochim. Acta* **332** 135455
[27] Hao J, Wang Y, Guo Q, Zhao J and Li Y 2019 Structural strategies for germanium-based anode materials to enhance lithium storage *Part. Part. Syst. Charact.* **36** 1900248
[28] Li D, Wang H, Liu H and Guo Z 2016 A new strategy for achieving a high performance anode for lithium ion batteries-encapsulating germanium nanoparticles in carbon nano-boxes *Adv. Energy Mater.* **6** 1501666
[29] Xue D, Xin S, Yan Y, Jiang K, Yin Y, Guo Y and Wan L 2012 Improving the electrode performance of Ge through Ge@C core–shell nanoparticles and graphene networks *JACS* **134** 2512–15
[30] Mo R, Rooney D, Sun K and Yang H 2017 3D nitrogen-doped graphene foam with encapsulated germanium/nitrogen-doped graphene yolk–shell nanoarchitecture for high-performance flexible Li-ion battery *Nat. Commun.* **8** 13949

[31] Zhang S, Zheng Y, Huang X, Hong J, Cao B, Hao J, Fan Q, Zhou T and Guo Z 2019 Structural engineering of hierarchical micro-nanostructured Ge–C framework by controlling the nucleation for ultralong-life Li storage *Adv. Energy Mater.* **9** 1900081

[32] Wang Y, Luo S, Chen M and Wu L 2020 Uniformly confined germanium quantum dots in 3D ordered porous carbon framework for high-performance Li-ion battery *Adv. Funct. Mater.* **30** 2000373

[33] Liu D, Liu Z, Li X, Xie W, Wang Q, Liu Q, Fu Y and He D 2017 Group IVA element (Si, Ge, Sn)-based alloying/dealloying anodes as negative electrodes for full-cell lithium-ion batteries *Small* **13** 1702000

[34] Kim H, Kim M, Choi S, Moon S, Kim Y and Park K 2019 Facile one-pot synthesis of Ge/TiO_2 nanocomposite structures with improved electrochemical performance *Nanoscale* **11** 17415–24

[35] Fang S, Shen L, Nie P, Xu G, Yang L, Zheng H and Zhang X 2015 Titanium dioxide/germanium core–shell nanorod arrays grown on carbon textiles as flexible electrodes for high density lithium-ion batteries *Part. Part. Syst. Charact.* **32** 364–72

[36] Hsieh M, Li G, Chang W and Tuan H 2017 A germanium nanoparticles/molybdenum disulphide (MoS_2) nanocomposite as a high-capacity, high-rate anode material for lithium-ion batteries *J. Mater. Chem.* A **5** 4114–21

[37] Li L, Wang D, Xu X, Guo X, Liu J, Mao C, Zhang Z and Li G 2021 In-situ construction of conducting alloy interphase towards modulating Li-ion storage kinetics *J. Energy Chem.* **56** 395–403

[38] Choe H, Kim M, Moon S, Kim E, Kim S, Lee G, Won J and Park K 2018 In-situ synthesis of Ge/Ti_4O_7 composite with enhanced electrochemical properties *Ceram. Int.* **44** 663–68

[39] Kim T, Park S, Lee T, Jeong J, Kim D, Swihart M, Song H, Kim J and Kim S 2017 ZnO decorated germanium nanoparticles as anode materials in Li-ion batteries *Nanotechnology* **28** 095402

[40] Song T *et al* 2014 Electrochemical properties of Si–Ge heterostructures as an anode material for lithium ion batteries *Adv. Funct. Mater.* **24** 1458–64

[41] Yang Y, Liu S, Bian X, Feng J, An Y and Yuan C 2018 Morphology- and porosity-tunable synthesis of 3D nanoporous SiGe alloy as a high-performance lithium-ion battery anode *ACS Nano* **12** 2900–8

[42] Zhu B *et al* 2019 Minimized lithium trapping by isovalent isomorphism for high initial Coulombic efficiency of silicon anodes *Sci. Adv.* **5** eaax0651

[43] Wang J, Du N, Zhang H, Yu J and Yang D 2012 Cu–Ge core–shell nanowire arrays as three-dimensional electrodes for high-rate capability lithium-ion batteries *J. Mater. Chem.* **22** 1511–15

[44] Wang X, Xu X, Liu J, Liu Z, Shen J, Li F, Hu R, Yang L, Ouyang L and Zhu M 2019 Facile synthesis of peapod-like Cu_3 Ge/Ge@C as a high-capacity and long-life anode for Li-ion batteries *Chemistry* **25** 11486–93

[45] Lin N, Zhou J, Han Y, Zhang K, Zhu Y and Qian Y 2015 Chemical synthesis of porous hierarchical Ge–Sn binary composites using metathesis reaction for rechargeable Li-ion batteries *Chem. Commun.* **51** 17156–59

[46] Kim D and Park C 2020 Co–Ge compounds and their electrochemical performance as high-performance Li-ion battery anodes *Mater. Today Energy* **18** 100530

[47] Yu J, Du N, Zhang H and Yang D 2013 Synthesis of Ni_xSi_y–SiGe core–shell nanowire arrays on Ni foam as a high-performance anode for Li-ion batteries *RSC Adv.* **3** 7713–17

[48] Yi S 2012 Graphene wrapped $LiFePO_4/C$ composites as cathode materials for Li-ion batteries with enhanced rate capability *J. Mater. Chem.* **22** 16465
[49] Wu J, Tang A, Huang S, Li J, Zeng L and Wei M 2020 *In situ* confined Co_5Ge_3 alloy nanoparticles in nitrogen-doped carbon nanotubes for boosting lithium storage *ACS Appl. Mater. Interfaces* **12** 46247–53
[50] Chang W, Lu S, Chu H and Tuan H 2018 Lithium-ion battery anodes of stacked nanowire laminate for ultrahigh areal capacities *ACS Sustain. Chem. Eng.* **7** 156–64
[51] Seng K, Park M, Guo Z, Liu H and Cho J 2013 Catalytic role of Ge in highly reversible GeO_2/Ge/C nanocomposite anode material for lithium batteries *Nano Lett.* **13** 1230–36
[52] Shang M, Chen X, Li B and Niu J 2020 A fast charge/discharge and wide-temperature battery with a germanium oxide layer on a Ti_3C_2 MXene matrix as anode *ACS Nano* **14** 3678–86
[53] Liu Y *et al* 2017 Local electric field facilitates high-performance Li-ion batteries *ACS Nano* **11** 8519–26
[54] Wei Y, Huang L, Chen J, Guo Y, Wang S, Li H and Zhai T 2019 Level the conversion/alloying voltage gap by grafting the endogenetic Sb_2Te_3 building block into layered GeTe to build $Ge_2Sb_2Te_5$ for Li-ion batteries *ACS Appl. Mater. Interfaces* **11** 41374–82
[55] Lv T, Li X and Ma J 2014 Hydrothermal synthesis of hollow $Ca_2Ge_7O_{16}$microspheres as high-capacity anodes for Li-ion batteries with long cycling life *RSC Adv.* **4** 49942–45
[56] Bauer D, Ashton T, Groves A, Dey A, Krishnamurthy S, Matsumi N and Darr J 2019 Continuous hydrothermal synthesis of metal germanates (M_2GeO_4; M¼ Co, Mn, Zn) for high-capacity negative electrodes in Li-ion batteries *Energy Technol.* **8** 1900692
[57] Li H, Wu X, Zhang L, Fan C, Wang H, Li X, Sun H, Zhang J and Yan Q 2016 Carbon-free porous Zn_2GeO_4 nanofibers as advanced anode materials for high-performance lithium ion batteries *ACS Appl. Mater. Interfaces* **8** 31722–28
[58] Haghighat Shishavan S, Nazarian Samani M, Nazarian Samani M, Roh H, Chung K, Oh S, Cho B, Kashani Bozorg S and Kim K 2019 Exceptionally reversible Li-/Na-ion storage and ultrastable solid-electrolyte interphase in layered GeP_5 anode *ACS Appl. Mater. Interfaces* **11** 32815–25
[59] Li W *et al* 2019 A self-healing layered GeP anode for high-performance Li-ion batteries enabled by low formation energy *Nano Energy* **61** 594–603
[60] Yi S 2015 Facile synthesis of porous V_2O_3/C composites as lithium storage material with enhanced capacity and good rate capability *J. Power Sources* **275** 392–98
[61] Kim Y, Hwang H, Lawler K, Martin S and Cho J 2008 Electrochemical behavior of Ge and GeX_2 (X = O, S) glasses: improved reversibility of the reaction of Li with Ge in a sulfide medium *Electrochim. Acta* **53** 5058–64
[62] Zhou X, Li T, Cui Y, Meyerson M, Weeks J, Mullins C, Jin Y, Liu Y and Zhu L 2020 *In situ* and operando morphology study of germanium–selenium alloy anode for lithium-ion batteries *ACS Appl. Energy Mater.* **3** 6115–20
[63] Nam K, Sung G, Choi J, Youn J, Jeon K and Park C 2019 New high-energy-density GeTe-based anodes for Li-ion batteries *J. Mater. Chem.* A **7** 3278–88
[64] Abel P, Chockla A, Lin Y, Holmberg V, Harris J, Korgel B, Heller A and Mullins C 2013 Nanostructured $Si_{(1-x)}Ge_x$ for tunable thin film lithium-ion battery anodes *ACS Nano* **7** 2249–57
[65] Chockla A, Klavetter K, Mullins C and Korgel B 2012 Solution-grown germanium nanowire anodes for lithium-ion batteries *ACS Appl. Mater. Interfaces* **4** 4658–64

[66] Klavetter K *et al* 2013 A high-rate germanium-particle slurry cast Li-ion anode with high Coulombic efficiency and long cycle life *J. Power Sources* **238** 123–36
[67] Chen J *et al* 2018 Intercalation of Bi nanoparticles into graphite results in an ultra-fast and ultra-stable anode material for sodium-ion batteries *Energy Environ. Sci.* **11** 1218–25
[68] Hwang J, Myung S and Sun Y 2017 Sodium-ion batteries: present and future *Chem. Soc. Rev.* **46** 3529–614
[69] Kim H, Kim H, Ding Z, Lee M, Lim K, Yoon G and Kang K 2016 Recent progress in electrode materials for sodium-ion batteries *Adv. Energy Mater.* **6** 1600943
[70] Chen W, Zhang X, Mi L, Liu C, Zhang J, Cui S, Feng X, Cao Y and Shen C 2019 High-performance flexible freestanding anode with hierarchical 3D carbon-networks/Fe_7S_8/graphene for applicable sodium-ion batteries *Adv. Mater.* **31** 1806664
[71] Song K, Liu C, Mi L, Chou S, Chen W and Shen C 2021 Recent progress on the alloy-based anode for sodium-ion batteries and potassium-ion batteries *Small* **17** 1903194
[72] Huang S, Liu L, Zheng Y, Wang Y, Kong D, Zhang Y, Shi Y, Zhang L, Schmidt O and Yang H 2018 Efficient sodium storage in rolled-up amorphous Si nanomembranes *Adv. Mater.* **30** 1706637
[73] Xu Y, Swaans E, Basak S, Zandbergen H, Borsa D and Mulder F 2016 Reversible Na-ion uptake in Si nanoparticles *Adv. Energy Mater.* **6** 1501436
[74] Kohandehghan A, Cui K, Kupsta M, Ding J, Memarzadeh Lotfabad E, Kalisvaart W and Mitlin D 2014 Activation with Li enables facile sodium storage in germanium *Nano Lett.* **14** 5873–82
[75] Abel P, Lin Y, de Souza T, Chou C, Gupta A, Goodenough J, Hwang G, Heller A and Mullins C 2013 Nanocolumnar germanium thin films as a high-rate sodium-ion battery anode material *J. Phys. Chem.* C **117** 18885–90
[76] Kim T, Song H and Kim S 2019 Production of germanium nanoparticles via laser pyrolysis for anode materials of lithium-ion batteries and sodium-ion batteries *Nanotechnology* **30** 275603
[77] Gavrilin I, Kudryashova Y, Kulova T, Skundin A and Gavrilov S 2021 The effect of growth temperature on the process of insertion/extraction of sodium into germanium nanowires formed by electrodeposition using indium nanoparticles *Mater. Lett.* **287** 129303
[78] Yi Z, Lin N, Li T, Han Y, Li Y and Qian Y 2019 Meso-porous amorphous Ge: Synthesis and mechanism of an anode material for Na and K storage *Nano Res.* **12** 1824–30
[79] Memarzadeh E, Kalisvaart W, Kohandehghan A, Zahiri B, Holt C and Mitlin D 2012 Silicon nanowire core aluminum shell coaxial nanocomposites for lithium ion battery anodes grown with and without a TiN interlayer *J. Mater. Chem.* **22** 6655–68
[80] Lotfabad E, Kalisvaart P, Kohandehghan A, Cui K, Kupsta M, Farbod B and Mitlin D 2014 Si nanotubes ALD coated with TiO_2, TiN or Al_2O_3 as high performance lithium ion battery anodes *J. Mater. Chem.* A **2** 2504–16
[81] Li Q, Zhang Z, Dong S, Li C, Ge X, Li Z, Ma J and Yin L 2017 Ge nanoparticles encapsulated in interconnected hollow carbon boxes as anodes for sodium ion and lithium ion batteries with enhanced electrochemical performance *Part. Part. Syst. Charact.* **34** 1600115
[82] Liu J, Muhammad S, Wei Z, Zhu J and Duan X 2020 Hierarchical N-doping germanium/carbon nanofibers as anode for high-performance lithium-ion and sodium-ion batteries *Nanotechnology* **31** 015402

[83] Wang X, Fan L, Gong D, Zhu J, Zhang Q and Lu B 2016 Core–shell Ge@Graphene@TIO_2 nanofibers as a high-capacity and cycle-stable anode for lithium and sodium ion battery *Adv. Funct. Mater.* **26** 1104–11
[84] Petrovičovà B *et al* 2021 Effect of germanium incorporation on the electrochemical performance of electrospun Fe_2O_3 nanofibers-based anodes in sodium-ion batteries *Appl. Sci.* **11** 1483–98
[85] Yue C *et al* 2015 High performance 3D Si/Ge nanorods array anode buffered by TiN/Ti interlayer for sodium-ion batteries *Adv. Funct. Mater.* **25** 1386–92
[86] Kajita T and Itoh T 2017 Ether-based solvents significantly improved electrochemical performance for Na-ion batteries with amorphous GeO_x anodes *Phys. Chem. Chem. Phys.* **19** 1003–9
[87] Fu L, Zheng X, Huang L, Shang C, Lu K, Zhang X, Wei B and Wang X 2018 Synthesis and investigation of $CuGeO_3$ nanowires as anode materials for advanced sodium-ion batteries *Nanoscale Res. Lett.* **13** 193
[88] Shen K, Lin N, Xu T, Han Y and Qian Y 2018 Amorphous mesoporous GeO $_x$ anode for Na-ion batteries with high capacity and long lifespan *Royal Soc. Open Sc.* **5** 171477
[89] Lim Y, Jung C, Im H, Park K, Park J, Cho W and Cha E 2016 Zn_2GeO_4 and Zn_2SnO_4 nanowires for high-capacity lithium- and sodium-ion batteries *J. Mater. Chem.* A **4** 10691–99
[90] Cui Y, Zhang R, Lei X and Liu X 2020 A stable tunnel-type $NaGe_{3/2}Mn_{1/2}O_4$ anode for Na-ion batteries *RSC Adv.* **10** 1426–29
[91] Li F, Qu Y and Zhao M 2016 Germanium sulfide nanosheet: a universal anode material for alkali metal ion batteries *J. Mater. Chem.* A **4** 8905–12
[92] Fu L, Shang C, Ma J, Zhang C, Zang X, Chai J, Li J and Cui G 2018 Cu_2GeS_3 derived ultrafine nanoparticles as high-performance anode for sodium ion battery *Sci. China Mater.* **61** 1177–84
[93] Li C, Wang B, Chen D, Gan L, Feng Y, Zhang Y, Yang Y, Geng H, Rui X and Yu Y 2020 Topotactic transformation synthesis of 2D ultrathin GeS_2 nanosheets toward high-rate and high-energy-density sodium-ion half/full batteries *ACS Nano* **14** 531–40
[94] Wang K, Liu M, Huang D, Li L, Feng K, Zhao L, Li J and Jiang F 2020 Rapid thermal deposited GeSe nanowires as a promising anode material for lithium-ion and sodium-ion batteries *J. Colloid Interface Sci.* **571** 387–97
[95] Kim D, Zhang K, Lim J, Lee G, Cho K, Cho M and Kang Y 2018 GeP_3 with soft and tunable bonding nature enabling highly reversible alloying with Na ions *Mater. Today Energy* **9** 126–36
[96] Li W, Li X, Liao J, Zhao B, Zhang L, Liu Y, Huang L, Li Y and Liu M 2019 Structural design of Ge-based anodes with chemical bonding for high-performance Na-ion batteries *Energy Storage Mater.* **20** 380–87
[97] Tseng K, Huang S, Chang W and Tuan H 2018 Synthesis of mesoporous germanium phosphide microspheres for high-performance lithium-ion and sodium-ion battery anodes *Chem. Mater.* **30** 4440–47
[98] Li X, Li W, Shen P, Yang L, Li Y, Shi Z and Zhang H 2019 Layered GeP-black P(Ge_2P_3): An advanced binary-phase anode for Li/Na-storage *Ceram. Int.* **45** 15711–14
[99] Zeng T, He H, Guan H, Yuan R, Liu X and Zhang C 2021 Tunable hollow nanoreactors for *in situ* synthesis of GeP electrodes towards high-performance sodium ion batteries *Angew. Chem. Int. Ed.* **60** 12103–8

[100] Khossossi N, Banerjee A, Essaoudi I, Ainane A, Jena P and Ahuja R 2021 Thermodynamics and kinetics of 2D g-GeC monolayer as an anode materials for Li/Na-ion batteries *J. Power Sources* **485** 229318
[101] Liu Y, Xiao X, Fan X, Li M, Zhang Y, Zhang W and Chen L 2018 GeP_5/C composite as anode material for high power sodium-ion batteries with exceptional capacity *J. Alloys Compd.* **744** 15–22
[102] Li W, Ke L, Wei Y, Guo S, Gan L, Li H, Zhai T and Zhou H 2017 Highly reversible sodium storage in a GeP_5/C composite anode with large capacity and low voltage *J. Mater. Chem.* A **5** 4413–20
[103] Suresh babu G and Kalaiselvi N 2021 Co_2GeO_4/graphene hetero-architecture as a potential anode for sodium ion batteries *J. Phys. Chem. Solids* **150** 109863
[104] Han J, Qin J, Guo L, Qin K, Zhao N, Shi C, Liu E, He F, Ma L and He C 2018 Ultrasmall Fe_2GeO_4 nanodots anchored on interconnected carbon nanosheets as high-performance anode materials for lithium and sodium ion batteries *Appl. Surf. Sci.* **427** 670–79
[105] Li M, Zhang Z, Ge X, Wei Z, Yao Y, Chen H, Wang C, Du F and Chen G 2018 Enhanced electrochemical properties of carbon coated Zn_2GeO_4 micron-rods as anode materials for sodium-ion batteries *Chem. Eng. J.* **331** 203–10
[106] Nam K, Jeon K and Park C 2019 Layered germanium phosphide-based anodes for high-performance lithium- and sodium-ion batteries *Energy Storage Mater.* **17** 78–87
[107] Shen H *et al* 2019 Sodium storage mechanism and electrochemical performance of layered GeP as anode for sodium ion batteries *J. Power Sources* **433** 126682
[108] Abel P, Fields M, Heller A and Mullins C 2014 Tin-germanium alloys as anode materials for sodium-ion batteries *ACS Appl. Mater. Interfaces* **6** 15860–67
[109] Farbod B, Cui K, Kalisvaart W, Martin M and Zahiri B 2014 Anodes for sodium ion batteries based on tin germanium antimony alloys *ACS Nano* **8** 4415–29
[110] Liu R *et al* 2021 Dual carbon decorated germanium–carbon composite as a stable anode for sodium/potassium-ion batteries *J. Colloid Interface Sci.* **584** 372–81
[111] Zhang W, Wu Z, Zhang J, Liu G, Yang N, Liu R, Pang W, Li W and Guo Z 2018 Unraveling the effect of salt chemistry on long-durability high-phosphorus-concentration anode for potassium ion batteries *Nano Energy* **53** 967–74

Chapter 4

Sn anode for various batteries

Zhijia Zhang, Mengmeng Zhang, Zhenyang Yu, Kai Sun, Shihao Sun, Heyi Sun, Gang Xie, Zhiheng Pang, Wenqiang Wang, Shenglong Yan, Yang Zhang, Pingfan Ning and Yong Jiang

Since the commercialization of Li-ion batteries in the late 1990s, Sn based anode has been one of the promising candidate materials for the next generation of Li-ion batteries due to its high theoretical capacity, suitable operating voltage and low cost. In addition, the lack of Li resources has promoted the revival of Na-ion batteries (SIBs) and K-ion batteries in recent years. Similar to Li metal, Sn can be electrochemically alloyed with Na or K, which has been widely and deeply studied. However, the electrochemical alloying/dealloying of Sn with Li/Na/K results in serious volume change, which prevents its commercial application. In order to solve this problem, researchers have proposed a variety of feasible solutions, such as reducing the particle size, constructing three-dimensional nanostructure, introducing buffer media and so on. After years of effort, the electrochemical performance of Sn based anode has obtained objectively improvement. In this chapter, the related research of Sn based anodes are reviewed, and the practical application progress of related types of batteries was prospected, which could provide some help for understanding the future development trend of high-performance rechargeable battery anode.

Abbreviations

Sn	Tin
Lithium	Li
Sodium	Na
Potassium	K
Carbon	C
Solid electrolyte interface	SEI
Chemical vapor deposition	CVD
Polymer C60	PC60

doi:10.1088/978-0-7503-4871-3ch4

Sn nanoparticles	SN NPs
Silicon dioxide	SiO_2
SnS nanoparticles	SnS NPs
SnS/C Nanofibers membrane	SnS/C NFM
SnSe nanoplates	SnSe/NC
Reduced graphene oxide	RGO
X-ray diffraction	XRD
End of discharge	EOD
End of charge	EOC
N-doped carbon nanofiber frameworks	Sn/N-CNFs
Sn nanoparticles restricted in carbon films	r-Sn/C
Pristine Sn	p-Sn
Nitrogen-doped porous carbon	Sn/NPC

4.1 Sn properties

In addition to C-based anode materials, other IVA group elements, such as Si, Ge and Sn, have been extensively studied due to their high theoretical capacity. By comparing the theoretical capacity and volume change of IVA group elements (figure 4.1(a)), Sn has higher theoretical specific capacity and its volume specific capacity is close to that of Si and Ge. In addition, Sn based anode materials have the advantages of high utilization, low cost and high conductivity. As early as 2005, amorphous Sn–Co–C composite anode material has been used in Sony's commercial Li-ion batteries, and its capacity is 30% higher than that of commercial Li-ion batteries at that time [1]. Therefore, Sn based anode has been widely concerned by researchers.

As the ability to form alloy with Li/Na/K ions, Sn can be used as anode of Li-ion batteries, Na-ion batteries and K-ion batteries, and can provide quite high specific capacity. The theoretical specific capacity of Sn is 994 mA h g^{-1} for Li-ion batteries according to $Li_{22}Sn_5$ [2], 847 mA h g^{-1} for Na-ion batteries according to $Na_{15}Sn_4$ [3], and 452 mA h g^{-1} for K-ion batteries according to K_2Sn [4]. However, during the Li/Na/K-ion insertion/extraction process, the volume of Sn based anode changes dramatically (Li-ion batteries: 260%, Na-ion batteries: 420%, K-ion batteries: 360%),

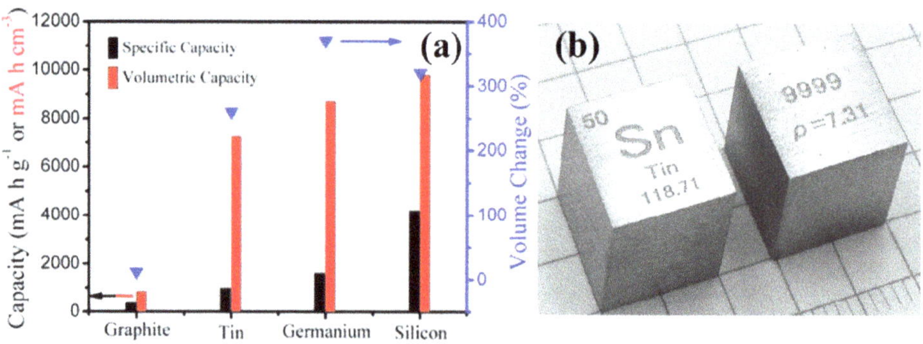

Figure 4.1. (a) Based on Li-ion batteries, the comparison of the capacity and volume changes for the IVA main group elements. Source: [1]. Reproduced with permission of John Wiley & Sons.

which leads to irreconcilable internal stress and a series of negative consequences: (a) the active material particles are crushed; (b) the collector loses electrical connection; (c) the newly formed surface continuously consumes Li/Na/K ions through the formation of SEImembrane; (d) particle aggregation leads to poor electrode dynamics. Currently, the problem of volume expansion cannot be completely eliminated. To solve this problem, Sn based anode modification is the main solution, such as reducing the particle size and introducing accommodation space or buffer to slow down the volume expansion, preparing of nanoporous structure for dispersing Sn in carbon matrix, and synthesizing Sn based alloys.

4.2 Development of Sn anode for Li, Na, K-ion batteries

Sn based materials have many common characteristics in synthesis and physico-chemical properties. This part introduces the preparation methods, characteristics and electrochemical properties of Sn based anode materials, and its electrochemical performances in Li-ion batteries, Na-ion batteries and K-ion batteries are also introduced. It is hoped that this summary could provide references for further research and practical application of Sn based materials.

4.2.1 Sn anode for Li-ion batteries

The theoretical capacity of Sn anode (994 mAh g^{-1}) is much higher than that of graphite (mAh g^{-1}) for Li-ion batteries. In addition, Sn anode combines lower cost and higher safety, which is more stable than graphite in commercial organic electrolyte. However, the practical application of Sn anode in Li-ion batteries is usually hindered by the serious volume change experienced in the process of lithium intercalation/extraction electrochemistry. These changes lead to cracks in the electrode structure, which quickly leads to the end of the cycle life. This problem can be controlled by appropriately modifying the electrode structure by reducing the particle size, combining with carbon material or selecting an optimized adhesive.

The nanosized dispersion of Sn could effectively alleviate the capacity attenuation caused by volume change during charging/discharging. Ryandaenggar *et al* chosen plasma assisted thermal evaporation combined with CVD to grow Sn nanoparticles in uniformly dispersed PC60 matrix [5]. The Sn-PC60 consists of 50 wt% Sn and 50 wt% PC60, which was used as the anode of the Li-ion battery. As shown in figures 4.2(a)–(c), Sn nanoparticles have appropriate spacing, and the particle size is less than 10 nm, which plays an important role in improving the diffusion and transport rate of lithium ion. PC60 has good electrochemical properties due to its relaxation and elasticity, which can buffer the volume change of tin ions in the reaction process. Under the current density of 1000 mAg^{-1}, the capacity is stable at 834 mAh g^{-1} for 5000 cycles, and the capacity retention rate reaches 97% (figure 4.2(d)).

The preparation of Sn based alloy is also an effective method to improve the electrochemical performance. Zhang *et al* [6] hydrogenated YSn_2 intermetallic compounds in a mechanical ball mill to obtain nanocomposite anode materials for Li-ion batteries, which were composed of Sn nanocrystals dispersed in the

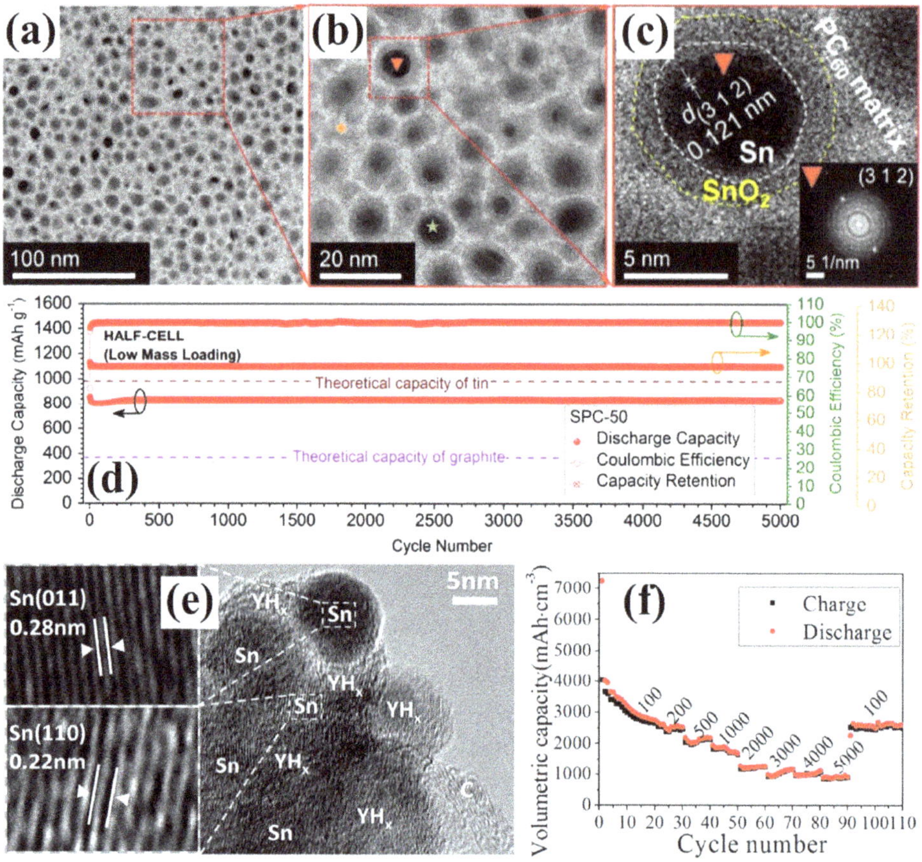

Figure 4.2. (a)–(c) TEM images of Sn-PC60; (d) Cycling ability of Sn-PC60 for 5000 cycles at 1000 mA g^{-1} of current density. Source: [5]. Reproduced with permission of American Chemical Society. (e) HRTEM image of Sn/YHx; (f) The rate performance of Sn/YHx. Source: [6]. Reproduced with permission of The Royal Society of Chemistry.

amorphous YH_X phase. As shown in figures 4.2(e) and (f), in the HRTEM image with high resolution, the clear lattice fringes with a distance of 0.28 nm and 0.22 nm match the (011) and (110) faces of Sn respectively, and the amorphous region corresponds to YH_X. This fully mixed structure can effectively reduce the volume expansion of tin ions in the process of lithiation/delithiation. When the current density increases to 5000 mAh g^{-1}, the capacity is also stable at 1000 mAh cm^{-3}. When the current density decreases to 100 mAh g^{-1} again, the capacity will return to the initial capacity, which has excellent rate performance.

Sn/metal oxide nanocomposites could also effectively improve the cycling stability of Sn based electrode. Kravchyk *et al* [7] prepared Sn/SnO_2 nanocomposites as anode materials for lithium-ion batteries by inorganic coating $SnCl_2$ and Lin, as shown in figure 4.3(a). It can be seen that uniform nanocrystalline network structure is obtained, and its average size is 9–23 nm, which can be adjusted. Electrochemical tests show that the Sn/SnO_2 composite with 10 nm has high electrochemical

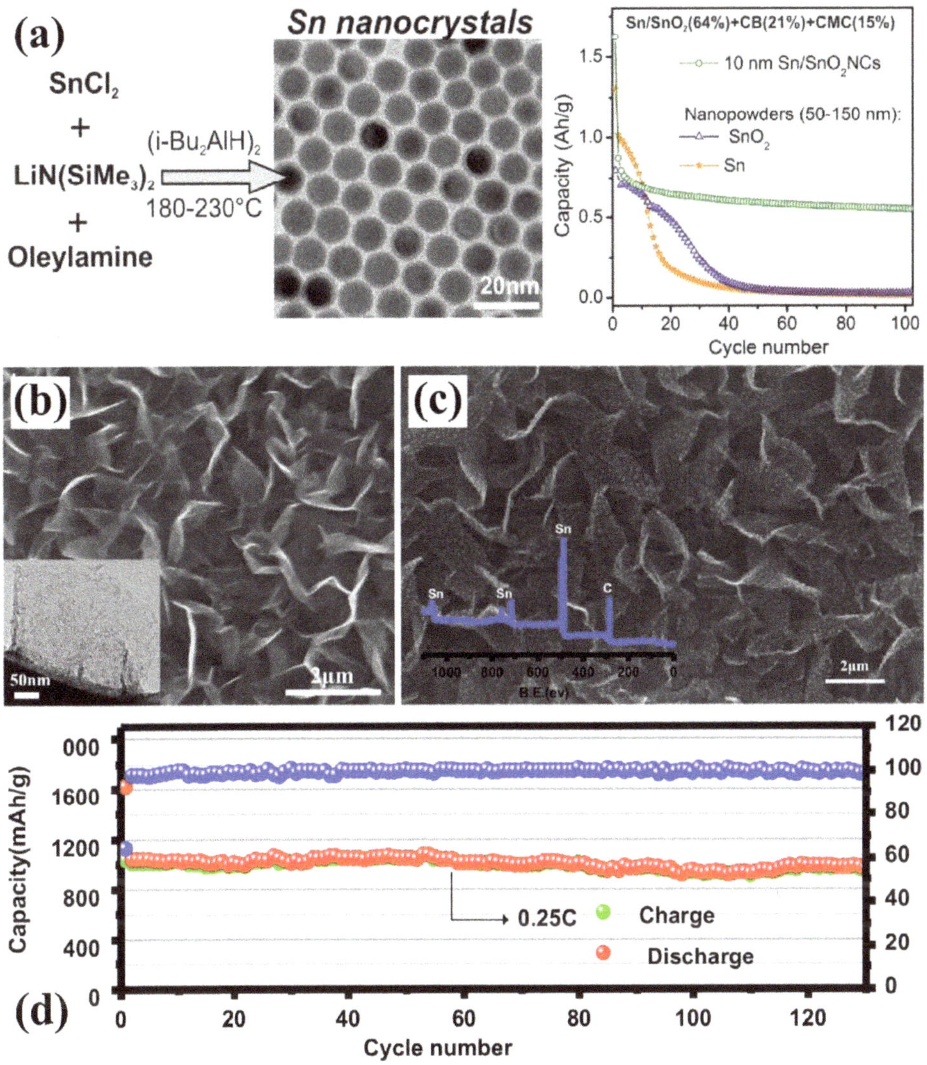

Figure 4.3. (a) Schematics of the reaction, TEM image for NCs, capacity and coulombic efficiency versus cycle number for active Sn based material at a current density of 1000 mAg^{-1}. Source: [7]. Reproduced with permission of American Chemical Society. (b) SEM images of SnO_2-VAGN. (c) SEM images of Sn@GB-VAGN. (d) Capacity and coulombic efficiency versus cycle number for SnO_2-VAGN at a current density of (0.25 C) 150 mAg^{-1}. Source: [8]. Reproduced with permission from Elsevier.

performance, especially after 100 deep charging (0.005–2 V) at 1000 mAh g^{-1} current, its reversible lithium storage capacity exceeds 700 mAh g^{-1}. Meanwhile, Li *et al* [8] used SnO_2 to encapsulate SN NPs in graphene nanostructures to prepare anode materials by microwave plasma irradiation. As shown in figures 4.3(b)–(d), Sn nanoparticles are encapsulated in graphene spheres and uniformly distributed on graphene nanosheets, which provides a larger surface area. The composite shows high capacity, long cycle life and excellent cycle stability. The capacity of the

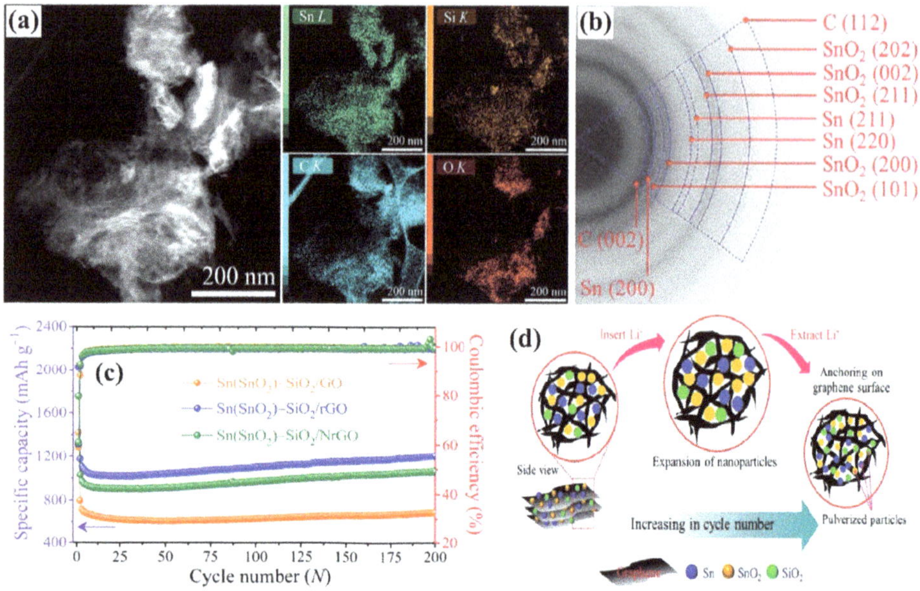

Figure 4.4. (a) High resolution TEM investigation and elemental maps of Sn, Si, C, and O for the Sn(SnO_2)–SiO_2/NrGO nanocomposite and the corresponding high-angle annular dark-field (HAADF) STEM image. (b) SAED patterns of Sn(SnO_2)–SiO_2/NrGO nanocomposite. (c) The above figures show the cycle performance of Sn(SnO_2)–SiO_2 graphene-based nanocomposites and (d) the simulation of crushing and buffering mechanisms in Sn(SnO_2)–SiO_2 graphene-based nanocomposites. Source: [9]. Reproduced with permission of The Royal Society of Chemistry.

composite can be maintained above 1000 mAh g^{-1} after 120 cycles at the current density of 150 mA g^{-1}.

Sn/SiO_2 nanocomposite with graphene sheets was synthesized by simple and low-cost technology [9]. The microstructure observation shows that the ultrafine Sn and SiO_2 nanoparticles are uniformly distributed on the graphene matrix (figures 4.4(a) and (b)). In addition, the phase of SnO_2 was observed in the composites. The electrochemical performance test shows that the composites prepared in the experiment have higher capacity compared with conventional graphite and graphene-based composites including Sn or SnO_2. The capacity of the composite can be maintained at 1205 mAh g^{-1} after 200 cycles (figures 4.4(c) and (d)). The excellent electrochemical properties of the new nanocomposite make them a good candidate material for high-performance anode materials of Li-ion batteries.

4.2.2 Sn anode for Na-ion batteries

Metal Sn is still regarded as a potential anode material with many advantages, such as low price, high theoretical capacity (847 mAh g^{-1}, $Na_{15}Sn_4$), environment-friendly and appropriate low charge–discharge potentials versus Na/Na^+. However, during the battery cycle, the transformation process from Sn to $Na_{15}Sn_4$ will cause serious volume expansion (about 420%). This huge volume fluctuation will lead to the crushing of electrode materials, making the cycle performance decline rapidly. Therefore, many techniques have been used to limit the volume expansion of Sn anode materials, inhibit the volume change and particle aggregation of Sn in the cycle process.

To solve the problems of Sn anode materials, such as powdering, falling off and low utilization, carbon nanocage is one of the most commonly be used to limit Sn volume fluctuation (figures 4.5(a), (d) and (g)) [10–12]. As figures 4.2(a)–(c) shown, Liu *et al* [10] embedded the ultrasmall Sn nanoparticles (≈ 8 nm) into spherical carbon network (8-Sn@C), and obtained the anode for SIBs with high rate performance, good stability and long cycle life. In addition, Ying *et al* [11] use a simple spray drying process to embed Sn nanodots in N-doped carbon (Sn/NMC) microcages uniformly

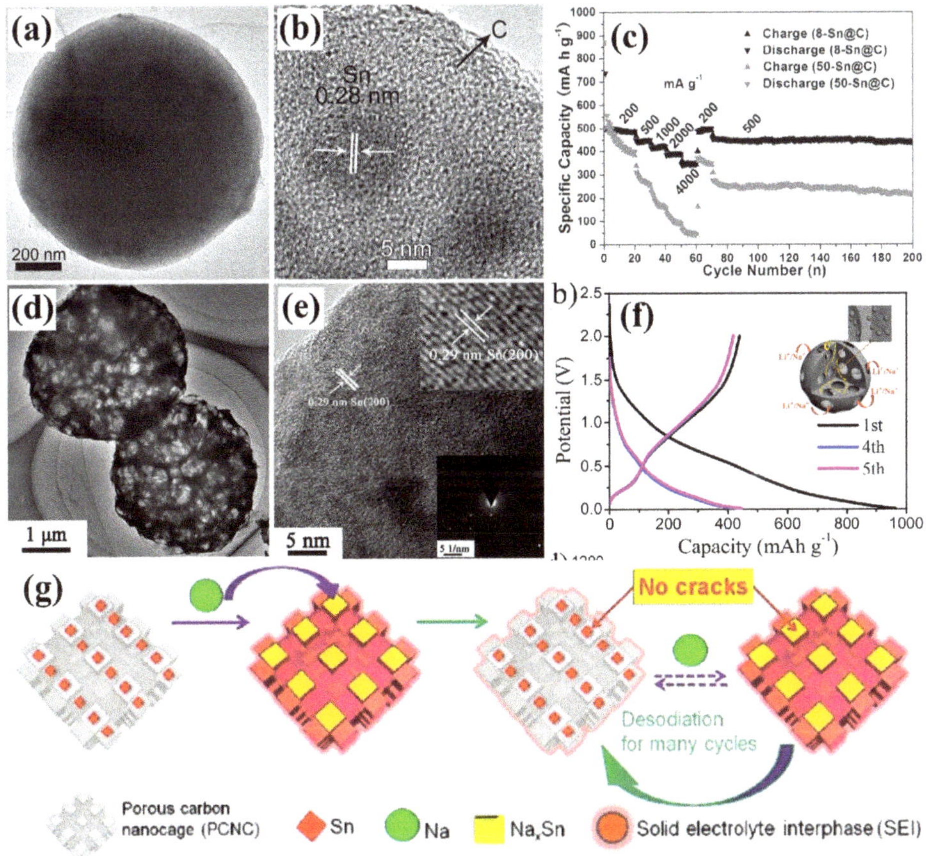

Figure 4.5. (a) TEM image of 8-Sn@C. (b) HRTEM images of 8-Sn@C nanocomposite after first charge. (c) Rate capability and cycling stability of the 8-Sn@C and 50-Sn@C electrodes with the range of voltage from 0.01 to 2 V. Source: [10]. Reproduced with permission of John Wiley & Sons. (d) TEM images of Sn/NMC. (e) HRTEM of Sn/NMC, (The lattice fringes patterns of Sn nanodots and SAED pattern of Sn/NMC were shown by insets). (f) Charge–discharge curves of Sn/NMC at 50 mA g^{-1}. Source: [11]. Reproduced with permission of The Royal Society of Chemistry. (g) Schematic illustration of suppressed volume expansion by the PCNCs during cycling process. Source: [12]. Reproduced with permission from Elsevier.

(figures 4.5(d) and (e)). This anode material can well limit the volume change of Sn, the initial reversible capacity of Sn/NMC is 439 mAh g^{-1} (figure 4.5(f)) with the current density of 50 mAh g^{-1} in SIBs and remains 332 mAh g^{-1} after 300 cycles.

Both selenide and sulfide of Sn have excellent theoretical capacities, which as anode material for Na-ion batteries have received considerable interest. Se and S belong to the same main group, so the selenide and sulfide of Sn have similar chemical properties in many aspects. SnSe/SnS shows a low electrical conductivity. And if it is directly used in Na-ion batteries would lead to low utilization of SnSe/SnS, which exhibits poor electrochemical performance. With the exploration of sulfides and selenides of Sn with different structures, it was found that SnSe/SnS with nano-structure or the composites combining SnSe/SnS and conductive materials as anodes were more likely to exhibit better rate capability and cycling stability for Na-ion batteries (figure 4.6).

In order to make use of the structural advantages of SnSe/SnS to be used as anode materials of Na-ion batteries, the usual method is to combine SnSe/SnS with carbon species to increase their conductivity and adapt to their volume changes. As shown in figures 4.6(a) and (b), Xiong *et al* [14] prepared SnS NPs and anchored it on 3D N-doped grapheme network by wet-chemical and freeze-drying processes with the advantages of simple. When using $Na_3V_2(PO_4)_3$/C as full Na-ion battery cathode, a high discharge capacity of 414.6 mAh g^{-1} could be obtained at 1 A g^{-1} after 300 cycles with voltage window from 0.8 to 3.0 V. Yan *et al* [15] used industrial SnO_2 as raw material to synthesize SnO_2/C nanofibers membrane (SnO_2/C NFM), which could be used to prepared SnS/C NFM by low-temperature vulcanization (figures 4.6(c)–(e)). This high conductivity and stable self-standing anode for Na-ion batteries can provide reversible capacity of 324 mAh g^{-1} after 200 cycles at 200 mA g^{-1} with a capacity retention rate of 84.8%. Ran *et al* [16] reported the SnSe/NC with strong Sn–C bonding were grown vertically on the nitrogen-doped carbon (figure 4.6(f)) and investigated its sodium storage mechanism(figures 4.6(g) and (h)), found that SnSe/NC has fast Na^+ diffusion ability and good cycling stability.

Wu *et al* prepared SnS–Sn–C nanocomposites by simple mechanical ball milling [13]. XRD and TEM results show that the introduction of S can reduce the size of Sn particles and form uniform SnS–Sn–C nanostructures. In the discharge process, Sn and Na_2S nanocrystals are generated by electrochemical conversion reaction of Sn particles, and then Sn particles are alloyed with Na to form NaSn alloy (figure 4.7). The conversion reaction of Sn and the alloying reaction of Sn and Na are completely reversible. SnS–Sn–C electrode showed a high capacity of 664 mAh g^{-1} and long-term cycling stability at 20 mA g^{-1}. These excellent electrochemical properties of SnS–Sn–C nanocomposites can be attributed to the continuous self-supporting matrix during the charge–discharge cycle, which provides a stable structure due to the introduction of S in the preparation process. Therefore, SnS–Sn–C nanocomposites are expected to be the anode materials for large capacity and low-cost energy storage systems of sodium-ion batteries.

In addition, the ordered nanostructures also play an important role in improving the electrochemical performance of Sn anode. Nam *et al* successfully synthesized Sn nanofibers with high aspect ratio from template free aqueous solution by a simple

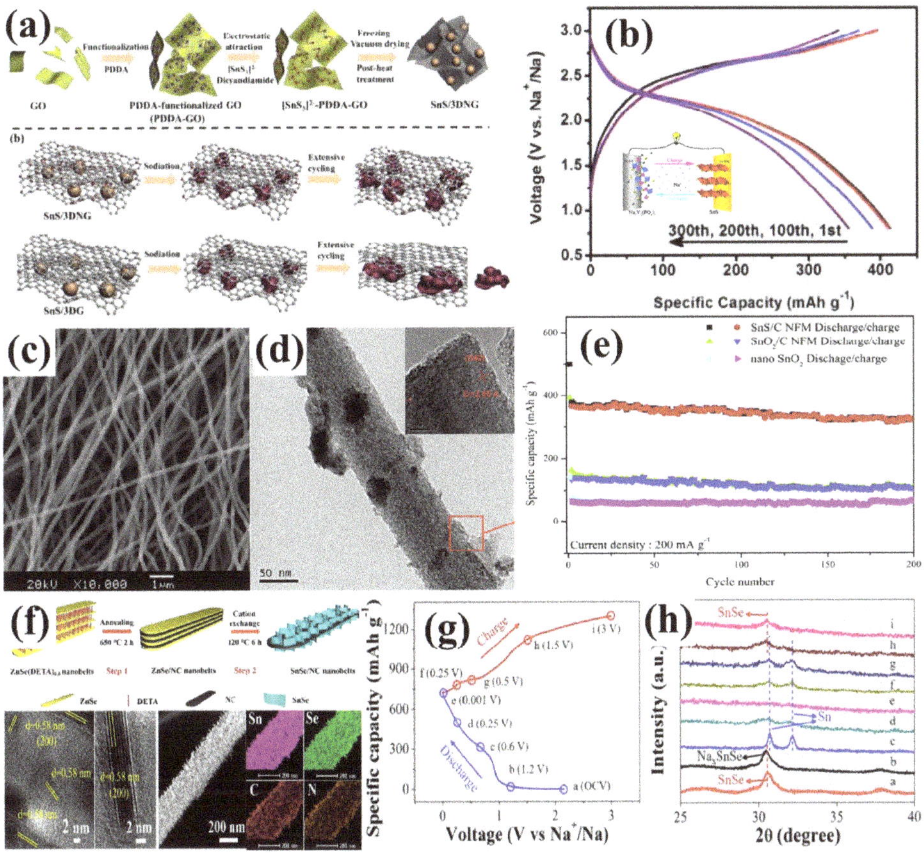

Figure 4.6. (a) Schematic illustration of the fabrication process of SnS/3DNG and the sodiation process in SnS/3DNG and SnS/3DG, respectively. (b) Charge–discharge curves of full Na-ion batteries at different cycles with the current density of 1 A g^{-1}. (c) FE–SEM images of SnS/C NFM. Source (a)–(c): [14]. Reproduced with permission of The Royal Society of Chemistry. (d) TEM image of the SnS/C NFM. (e) Cycling capability of SnS/C NFM, SnO_2/C NFM and nano SnO_2 with the current density of 200 mA g^{-1}. (f) Schematic illustration of SnSe/NC hybrid nanobelts preparation process and HRTEM image of SnSe/NC: HAADF–STEM image and energy dispersive spectrum of x-ray of SnSe/NC. Source (d)–(f): [15]. Reproduced with permission from Elsevier. (g) Charge–discharge curves and (h) *Ex situ* XRD patterns of SnSe/NC electrode at the first cycle. Source (g)–(h): [16]. Reproduced with permission from Elsevier.

electrodeposition [17]. This method involves the rapid electrochemical deposition of Sn and the strong adsorption of Triton X-100 on the {200} plane of Sn grains, thus inducing the anisotropic growth of one-dimensional Sn nanofibers perpendicular to the (112) or (1 12) plane (figure 4.8). As electrode materials for sodium-ion batteries, Sn nanofibers exhibit high reversible capacity and excellent cycling performance. After 100 cycles, the charge capacity remains at 776.26 mAh g^{-1}, which is equivalent to 95.09% of the initial charge capacity. The excellent electrochemical properties of Sn nanofibers are mainly attributed to their high mechanical stability to sodium, which is due to their highly anisotropic expansion and the pore volume between the nanofibers. Although further research is needed to improve the coulomb efficiency

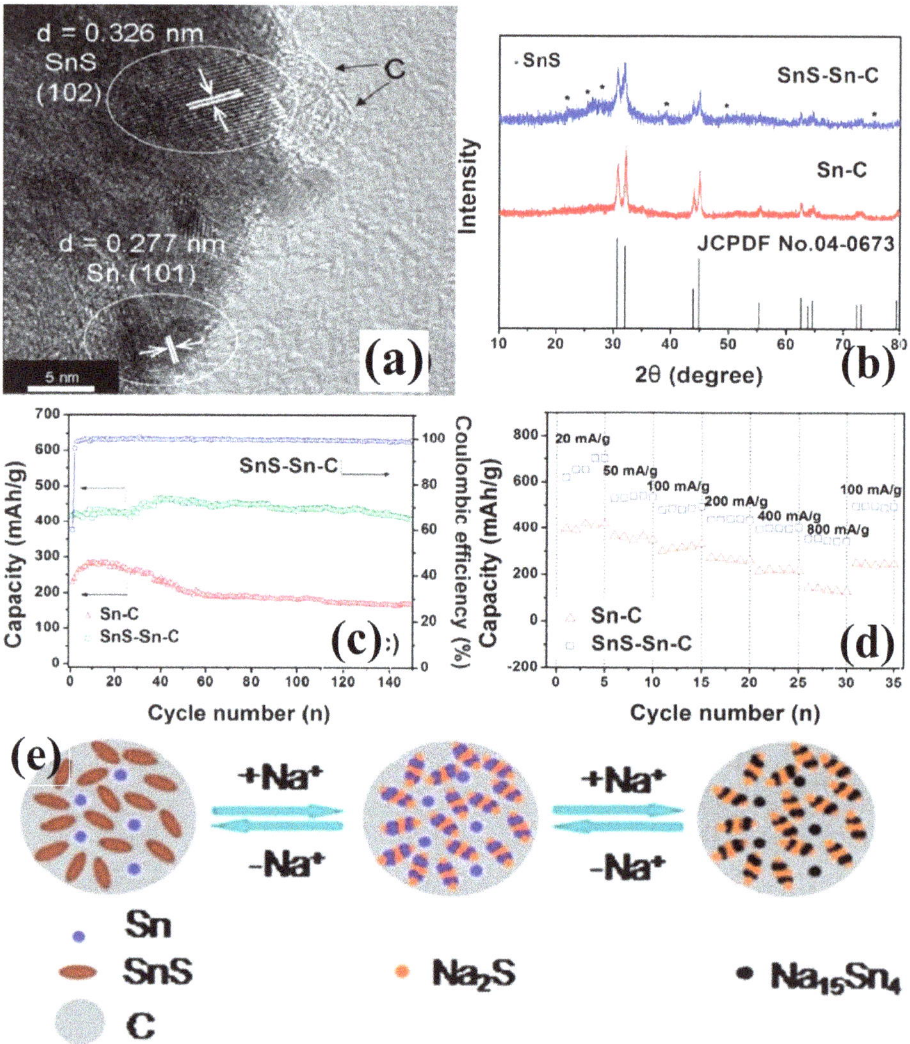

Figure 4.7. (a) High resolution TEM image of the SnS–Sn–C nanocomposite. (b)XRD patterns of nanosized SnS–Sn–C and Sn–C nanocomposites. (c) Cycling performance of the SnS–Sn–C and Sn–C electrodes at a cycling rate of 100 mAg^{-1}. (d) Rate capability of the SnS–Sn–C and Sn–C electrodes at different current rates from 20 to 800 mA g^{-1}. (e) Schematic diagram of sodium insertion/removal for SNS–Sn–C anode. Source: [13]. Reproduced with permission of The Royal Society of Chemistry.

and rate capability, tin nanofibers synthesized by this simple electrodeposition process prove the feasibility of using binderless and conductive tin as anode materials for high-performance Na-ion batteries.

4.2.3 Sn anode for K-ion batteries

K-ion battery has been considered as an excellent substitute for lithium-ion battery because of its low price and low potential. Due to the short research time of K-ion

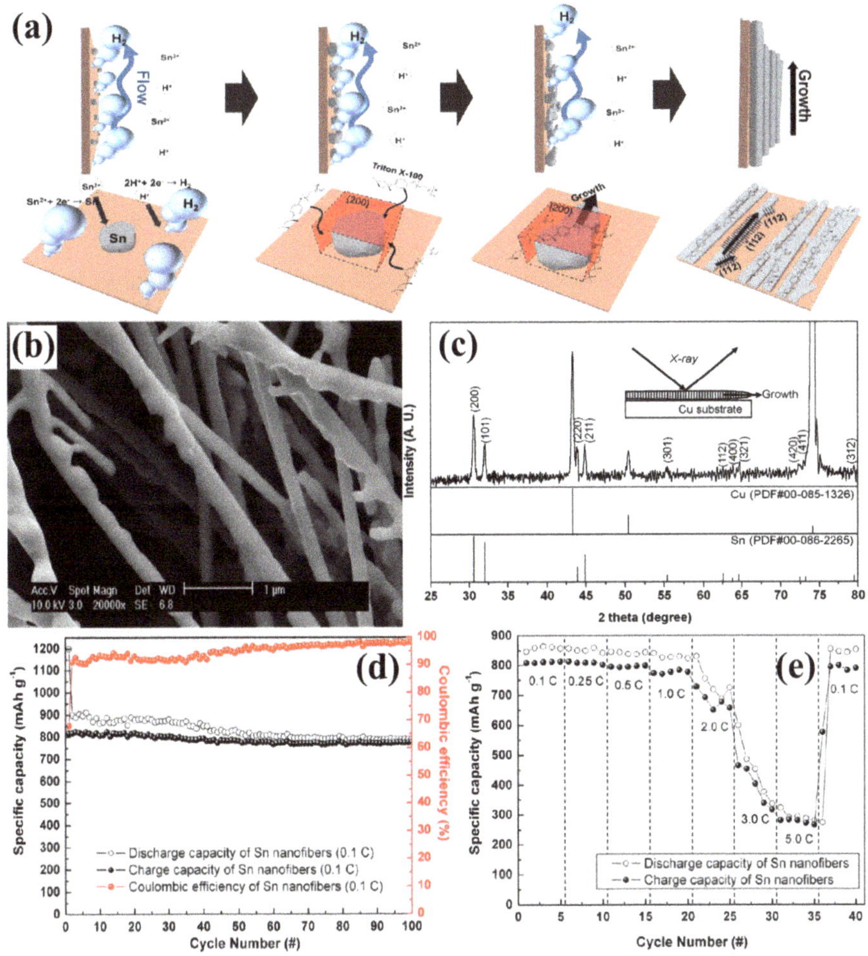

Figure 4.8. (a) Schematic diagram of the growth mechanism of the Sn nanofibers by cathodic electrodeposition. (b) SEM images of the Sn nanofibers synthesized by cathodic electrodeposition at 10.5 V for 30 s were obtained from an aqueous solution containing 0.25 M $SnSO_4$, 0.7 M H_2SO_4 and 40 g l^{-1} Triton X-100. (c) XRD patterns of the Sn nanofibers synthesized by cathodic electrodeposition at 10.5 V for30 s from an aqueous solution containing 0.25 M $SnSO_4$, 0.7 M H_2SO_4 and 40 g l^{-1} Triton X-100. (d) Cycle performance and Coulombic efficiency of the Sn nanofibers at a rate of 0.1 Cover a voltage range of 0.001 to 0.65 V (vs Na/Na^+). (e) Specific capacities of the Sn nanofibers with the current density varying from 0.1 to 5 C and back to0.1 C between 0.001 and 0.65 V with a fixed discharging current density of 0.1 C. Source: [17]. Reproduced with permission of American Chemical Society.

batteries, however, it is still difficult for Sn based anode materials to achieve high capacity and long cycle life in K-ion batteries. Therefore, it is necessary to study Sn based anode materials with high capacity and long life [18].

Sn based anode materials have excellent performance in lithium/sodium-ion batteries. Sn and K in different atomic ratio can form a variety of different binary alloys.

The higher density of Sn can lead to higher bulk energy density, which indicates that Sn and Sn based materials have good potential in K-ion batteries as other alloy anode materials. However, serious volume expansion (up to 360%) occurs during the charging/discharging process. It will cause the electrode material broken and broken, and is not conducive to the stability of the protective layer between the electrolyte and the negative electrode, which will have a great impact on the battery life and rechargeable performance. Therefore, we need to find a good way to alleviate its volume expansion.

Wang and colleagues proposed to use reduced graphene oxide network encapsulated nano submicron particles as K storage anode (figure 4.9). The XRD patterns of the device were analyzed Sn@RGO. The results show that tin phase crystallizes well in the composites. At the same time, the weak peak and shoulder peak of 26.1°are due to the addition of RGO material, which indicates that the d-spacing of RGO network layer is larger than graphite (002) plane. Because of the addition of porous RGO network, not only the specific surface area of the material is increased, but also the conductivity of the material is improved. Therefore, the capacity of 200 mAh g^{-1} can be maintained after 50 cycles, which is much higher than that of pure Sn electrode (12.9 mAh g^{-1}). The results show that changing the microstructure of Sn based anode material has a great influence on its electrochemical performance. The composite also shows that tin based anode material has excellent performance in K-ion battery.

Sn based alloy materials as anode electrodes of K-ion batteries are systematically studied. The morphology evolution of Sn and SnSb electrodes after cycling is shown in figure 4.10 [20]. The SEM images of the original Sn and SnSb electrodes show that

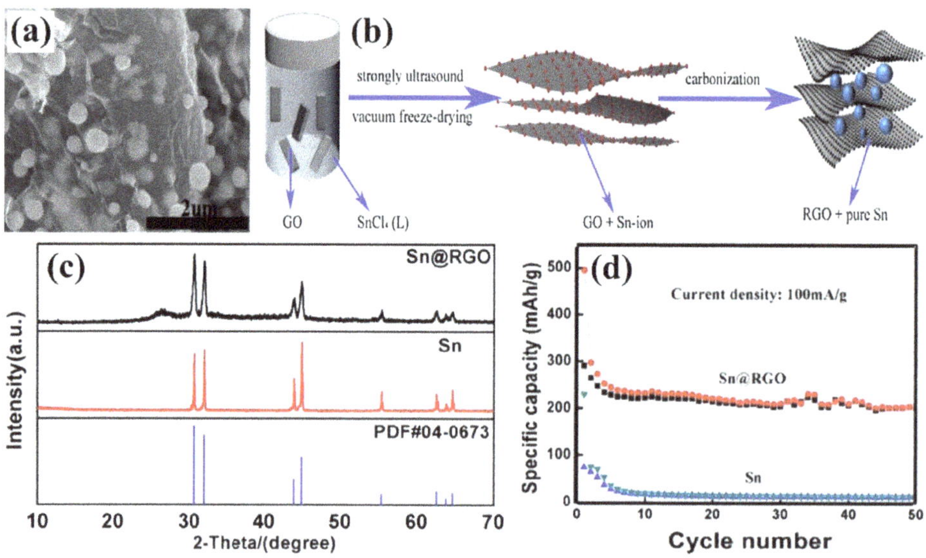

Figure 4.9. (a) Sn@RGO.SEM images; (b) Sn@RGO Synthesis process diagram; (c) Pure Sn and Sn@RGO XRD of electrode; (d) Cycling performance Sn and Sn@RGO electrodes at 100 mAg^{-1}. Source: [19]. Reproduced with permission from American Chemical Society.

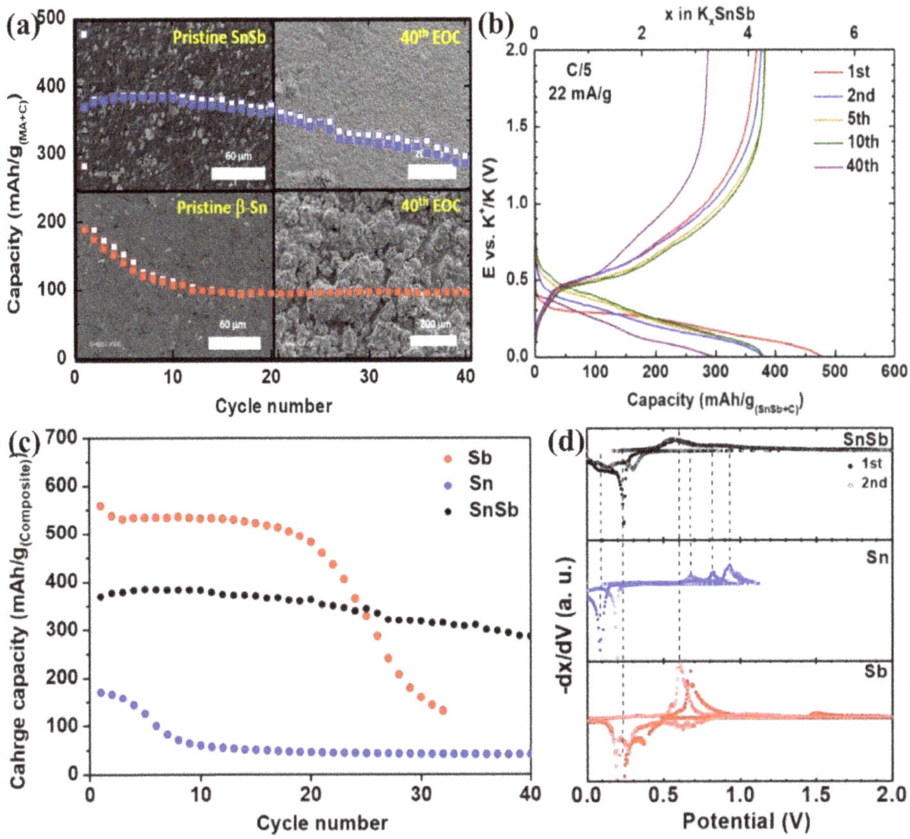

Figure 4.10. (a) Cycling performance of SnSb and Sn electrodes in K half-cells and SEM images of SnSb (upper) and Sn (lower) electrodes. (b) Galvanostatic discharge/charge profile of SnSb electrode. (c) Cycling performance of SnSb, Sn and Sb electrodes at C/5 rate in K half-cells. (d) Derivatives of the first cycle of the galvanostatic profiles. Source: [20]. Reproduced with permission from The Royal Society of Chemistry.

the active particles are evenly distributed (figure 4.10(a)). At the EOD, the surface of the electrode looks very different during explosive disposal. The Sn electrode reveals irregular large particles embedded in non-uniform SEI layer, while SnSb electrode reveals uniform particles and SEI coverage.

At the EOC, because of volume of Sn electrode decreases during discharge, many cracks are observed on the Sn surface. No cracks are found on the surface of SnSb electrode, where the particles are evenly distributed (figure 4.10(a)). During the cycling process, the morphology of SnSb electrode does not change, and there is no crack in the submicron range, which explains that alloying/dealloying can improve the reversibility of the reaction when using intermetallic compound SnSb instead of Sn. Figures 4.10(b)–(d) show the capacity retention of SnSb electrode compared with pure Sn and Sb electrode. The capacity of SnSb electrode decreases gradually and maintains 75% initial reversible capacity after 40 cycles, while the capacity of Sn electrode decreases from the first cycle of Sn and the capacity of Sb electrode decreases from the 20th cycle of Sb. These results provide a new idea for the design of anode alloy materials for K-ion batteries (figures 4.10(b)–(d)). In fact, the progressive potassium evolution of SnSb during the formation of amorphous phase allows the integrity of the electrode to be maintained during cycling, despite significant volume expansion.

Sn/C composites have been widely and deeply studied as anode materials. Chao *et al* synthesized a special hybrid material Sn/N-CNFS-5 through electrospinning and subsequent carbonization process, which combines the advantages of porous Sn nanospheres composed of ultrafine nanoparticles with the advantages of staggered N-CNF skeleton with high structural integrity (figure 4.11) [21]. The porous Sn nanospheres provide enough void space to adapt to the volume change during the cycle, provide more active centers for K^+ storage, and shorten the diffusion distance of K^+(figure 4.11(c)). The amorphous carbon skeleton prevents the internal Sn nanoparticles from accumulating in the cycling process (figures 4.11(d) and (e)), and N-doped carbon nanoparticles improve the stability and electrical conductivity of the electrode (figures 4.11(f)–(i)). The results show that SN/N-CNFS-5 has an extremely long cycle life (198.0 mA h g^{-1} at 1 A g^{-1}, after 3000 cycles, capacity retention up to 88.4%) (figure 4.11 (b)).

Meng *et al* refined Sn particles from submicron (300 nm) to nanometer (88 nm) by oxidize–reduction treatment (figures 4.12(a)–(d)) [22]. Through a series of experiments, they confirmed that the refinement of Sn follows a decentralized nucleation

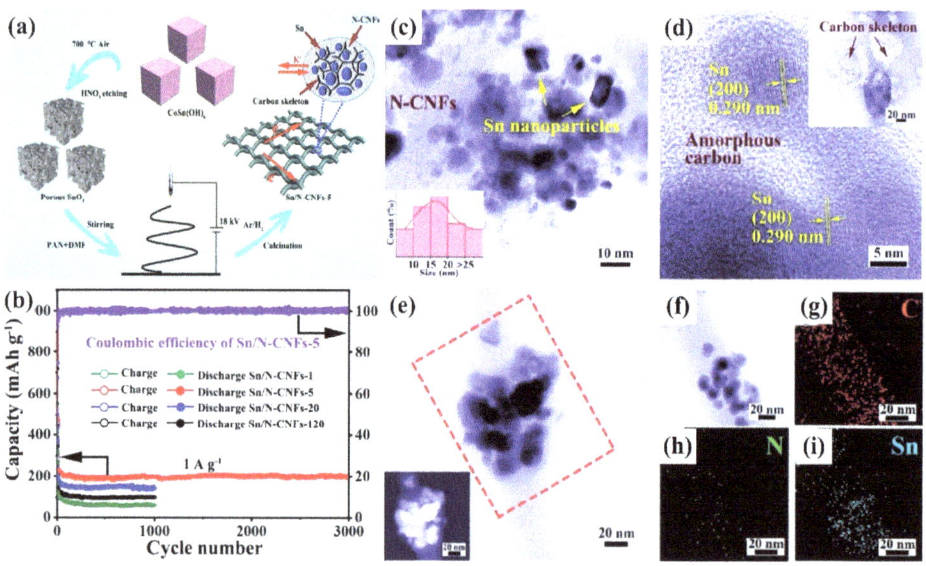

Figure 4.11. (a) Schematic illustration of the preparation procedure of Sn/N-CNFs-5. (b) Cycling performances and CE of Sn/N-CNFs at 1 A g^{-1}. (c) TEM image of grain size distribution of Sn/N-CNFS-5 (inset). (d) HRTEM images of Sn/N-CNFs-5. (d) The illustration shows the HF etched TEM image of SN/N-CNFS-5. (e) TEM image and HAADF–STEM image corresponding to SN/N-CNFS-5 (inset). (f) TEM image of Sn/N-CNFs-5. (g)–(i) The mapping image of the corresponding elements of C, N, and Sn. Source: [21]. Reproduced with permission of The Royal Society of Chemistry.

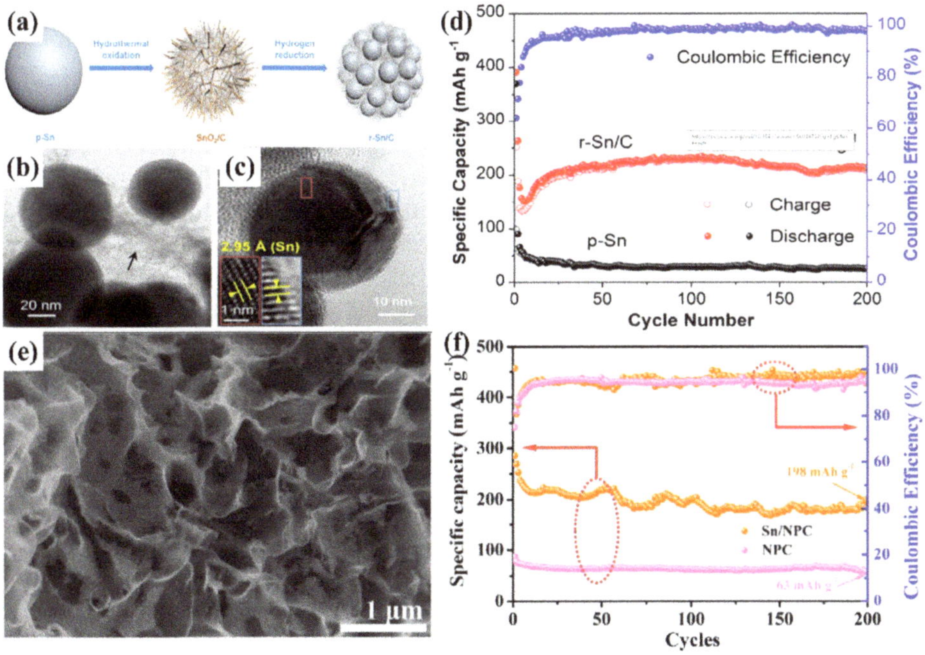

Figure 4.12. (a) The 'oxidation–reduction' process from p-Sn to SnO_2/C and then to R-Sn/C. (b) Magnified TEM image. (c) Enlarged TEM images of individual nanoparticles with a high resolution TEM image in the illustration. (d) The cyclic performance of R-Sn/C and p-Sn at the current density of 0.1 A g^{-1} is related to the coulomb efficiency of R-Sn/C. Source (a)–(d): [22]. Reproduced with permission of American Chemical Society. (e) High power scanning electron microscope image of SN/NPC. (f) Cyclic properties of Sn/NPC and NPC anodes at 50 mA g^{-1}. Source: [23]. Reproduced with permission from Elsevier.

and restricted fusion mechanism. The carbon film formed in the oxidation process inhibits the agglomeration of tin nanoparticles and maintains a narrow particle size distribution, which plays a key role in grain refinement. As the anode of K-ion battery, the obtained tin nanoparticles are confined in the carbon film (R-Sn/C), showing better cycle stability and capacity retention (figure 4.12(d)).

Yang *et al* [23] prepared Sn nanoparticles immobilized uniformly on Sn/NPC by a simple sol–gel method and heat treatment (figure 4.12(e)). NPC, as the carrier framework, gives good conductivity to Sn/NPC and prevents agglomeration of Sn nanoparticles, which is conducive to improving the electrochemical performance. As an anode for K-ion batteries, the capacity can maintain at 198 mAh g^{-1} at the current density of 50 mA g^{-1} after 200 cycles (figure 4.12(f)).

4.3 Challenge of Sn anode for Li, Na, K-ion batteries

In future research work, controlling volume change, aggregation of nanoparticles and excessive consumption of SEI to active materials and electrolyte are still the main research directions of Sn based anode materials. In terms of practical application, Sn/C composite has the largest commercial application prospect. Carbon nanostructures can effectively prevent the powdering and re-aggregation of the active materials, thus reducing the influence of the volume change of Sn based anode and maintaining stable electrical contact. In addition, carbon nanostructures can effectively protect the active materials from direct contact with the electrolyte, thus inhibiting the formation of SEI films. At the same time, Sn/C composite is suitable for practical application and promotion in both synthesis process and production cost. In terms of basic research, researchers will continue to pay attention to the construction of new micro/nanostructures. Because Sn based anode materials with excellent structure can effectively improve the electrochemical performance and optimize the mechanism of metal ions insertion/de-insertion. However, simplifying the preparation process and reducing the cost are always the only way from basic research to practical application. In addition to nano-scale regulation, micro-sized materials may become a new trend in Sn based anode research. The coulombic efficiency and energy density of electrode can be effectively improved by adjusting the structure at micro-scale, but its cycle stability is still facing the main problem.

After systematic modification, great progress has been made in improving the electrochemical performance of Sn based anode materials. There is still no perfect solution, however, for the key problems such as pulverization and Li/Na/K source loss. Finally, the practical application of Sn based anode needs to be optimized by batch preparation process. Sn based materials have many challenges in the field of electrochemical energy storage, but it is undeniable that they have great potential in high-performance energy storage systems. It is hoped that the contents of this chapter can provide some useful reference and help for the research and practical application of Sn based anode materials.

References

[1] Ying H and Han W-Q 2017 Metallic Sn-based anode materials: application in high-performance lithium-ion and sodium-ion batteries *Adv. Sci.* **4** 1700298

[2] Qin J, He C, Zhao N, Wang Z, Shi C, Liu E-Z and Li J 2014 Graphene networks anchored with Sn@graphene as lithium ion battery anode *ACS Nano.* **8** 1728–38

[3] Choi Y-S, Byeon Y-W, Ahn J-P and Lee J-C 2017 Formation of zintl ions and their configurational change during sodiation in Na–Sn battery *Nano Lett.* **17** 679–86

[4] Pramudita J C, Sehrawat D, Goonetilleke D and Sharma N 2017 An initial review of the status of electrode materials for potassium-ion batteries *Adv. Energy Mater.* **7** 1602911

[5] Ryanda Enggar Anugrah A, Guicheng L, Minh Xuan T, Chairul H, Ji Young K, Hyunjin Y and Joong Kee L 2018 Self-relaxant superelastic matrix derived from C_{60} incorporated Sn nanoparticles for ultra-high-performance Li-ion batteries *ACS Nano.* **12** 5588–604

[6] Xinyao Z, Chengkai Y, Xinghua C, Zhiliang L, Xiuqi W, Henghui Z, Jie Z and Xingguo L 2018 A high capacity nanocrystalline Sn anode for lithium ion batteries from hydrogenation induced phase segregation of bulk ysn2† *J. Mater. Chem.* A **6** 21266–73

[7] Kostiantyn K, Loredana P, Maryna I B, Frank K, Maksym Y, Marc W, Christoph G and Maksym V K 2013 Monodisperse and inorganically capped Sn and Sn/SnO_2 nanocrystals for high-performance Li-ion battery anodes *JACS* **135** 4199–202

[8] Na L, Huawei S, Hao C and Chengxin W 2013 Sn@graphene grown on vertically aligned graphene for high-capacity, high-rate, and long-life lithium storage *Nano Energy.* **3** 102-12

[9] Namsar O, Autthawong T, Laokawee V, Boonprachai R, Haruta M, Kurata H, Yu A, Chairuangsri T and Sarakonsri T 2020 Improved electrochemical performance of anode materials for high energy density lithium-ion batteries through $Sn(SnO_2)$–SiO_2/graphene-based nanocomposites prepared by a facile and low-cost approach. *Sustain. Energy Fuels.* **4** 4625–36
[10] Liu Y, Zhang N, Jiao L, Tao Z and Jun J 2014 Ultrasmall sn nanoparticles embedded in carbon as high-performance anode for sodium-ion batteries *Adv. Funct. Mater.* **25** 214–20
[11] Ying H, Zhang S, Meng Z, Sun Z and Han W-Q 2017 Ultrasmall sn nanodots embedded inside N-doped carbon microcages as high-performance lithium and sodium ion battery anodes *J. Mater. Chem.* A **5** 8334–42
[12] Chen S, Ao Z, Sun B, Xie X and Wang G 2016 Porous carbon nanocages encapsulated with tin nanoparticles for high performance sodium-ion batteries *Energy Storage Mater.* **5** 180–90
[13] Wu L, Hu X, Qian J, Pei F, Wu F, Mao R, Ai X, Yang H and Cao Y 2013 A Sn–SnS–C nanocomposite as anode host materials for Na-ion batteries *J. Mater. Chem.* A **1** 7181–84
[14] Xiong X, Yang C, Wang G, Lin Y, Ou X, Wang J-H, Zhao B, Liu M, Lin Z and Huang K 2017 SnS nanoparticles electrostatically anchored on three-dimensional N-doped graphene as an active and durable anode for sodium-ion batteries *Energy Environ. Sci.* **10** 1757–63
[15] Yan H, Yang M, Liu L, Xia J, Yuan Y, Liu J, Zhang Y, Nie S and Wang X 2020 Synthesis of SnS/C nanofibers membrane as self-standing anode for high-performance sodium-ion batteries by a smart process *J. Alloys Compd.* **843** 155899
[16] Ren X, Wang J, Zhu D, Li Q, Tian W, Wang L, Zhang J, Miao L, Chu P K and Huo K 2018 Sn–C bonding riveted snse nanoplates vertically grown on nitrogen-doped carbon nanobelts for high-performance sodium-ion battery anodes *Nano Energy.* **54** 322–30
[17] Nam D-H, Kim T-H, Hong K-S and Kwon H-S 2014 Template-free electrochemical synthesis of Sn nanofibers as high-performance anode materials for Na-ion batteries *ACS Nano.* **8** 11824–35
[18] Yang Y, Li D, Zhang J, Suo G, Yu Q, Feng L, Hou X, Ye X, Zhang L and Wang W 2019 Sn nanoparticles anchored on N doped porous carbon as an anode for potassium ion batteries *Mater. Lett.* **256** 126613
[19] Wang H, Xing Z, Hu Z, Zhang Y, Hu Y, Sun Y, Ju Z and Zhuang Q 2019 Sn-based submicron-particles encapsulated in porous reduced graphene oxide network: advanced anodes for high-rate and long life potassium-ion batteries *Appl. Mater. Today* **15** 58–66
[20] Gabaudan V, Berthelot R, Sougrati M T, Lippens P-E, Monconduit L and Stievano L 2019 SnSb vs. Sn: improving the performance of Sn-based anodes for K-ion batteries by synergetic alloying with Sb *J. Mater. Chem.* A **7** 15262–70
[21] Chao L, An Tong B, Hong Li C, Ya Ru P, Ming Z, Chun Cheng Y and Qing J 2021 Rational design of porous Sn nanospheres/N-doped carbon nanofibers as an ultra-stable potassium-ion battery anode material *J. Mater. Chem.* A **9** 5740–50
[22] Meng Y, Lie Y, Liuyang S, Yongqiang Z, Narayan R, Yuanbin Q, Qinqin F, Liqiang Z and Xiaohui N 2021 Refined tin nanoparticles by oxidation–reduction treatment for use in potassium-ion batteries *ACS Appl. Nano Mater.* **4** 4432–40
[23] Yanling Y, Dan L, Jiaqi Z, Guoquan S, Qiyao Y, Lei F, Xiaojiang H, Xiaohui Y, Li Z and Wei W 2019 Sn nanoparticles anchored on n doped porous carbon as an anode for potassium ion batteries *Mater. Lett.* **256** 126613

IOP Publishing

Metal Electrodes for Battery Technologies

Yurong Ren and Jianmin Ma

Chapter 5

Antimony as the negative electrode for various batteries

Lei Zhang

Lithium-ion batteries (LIBs) are the most successful energy storage system since the commercialization in the early 1990s. Unfortunately, the low theoretical capacity of the commercialized graphite-based anode is still the critical issue for further improvement of the energy density in LIBs. Therefore, the antimony (Sb)-based anodes with high theoretical capacities have attracted tremendous research attention as promising candidates for next-generation LIBs. In this chapter, we summarize the most promising Sb-based anodes and their applications in three different energy storage systems, such as LIBs, sodium-ion batteries (SIBs) and potassium-ion batteries (PIBs). Additionally, various strategies for enhancing the battery performance of Sb-based anodes are presented, including the carbon coating, graphene introduction, porous structure design etc. Moreover, an outlook on the further improvement of Sb-based anode materials is also presented.

Abbreviations

LIBs	Lithium-ion batteries
SIBs	Sodium-ion batteries
PIBs	Potassium-ion batteries
Sb	Antimony
NPs	Nanoparticles
XRD	X-ray diffraction
XPS	X-ray photoelectron spectrum
TEM	Transmission electron microscopy
HRTEM	High-resolution transmission electron microscopy
SEM	Scanning electron microscopy
DFT	Density functional theory
rGO	Reduced graphene oxide
CNT	Carbon nanotubes

doi:10.1088/978-0-7503-4871-3ch5

SEI	Solid electrolyte interface
ICE	Initial coulombic efficiency
1D	One dimensional
3D	Three dimensional

5.1 Sb properties

Antimony (Sb) is one of the most promising anode materials for energy storage systems. The reversible electrochemically reactions can be happened between Sb and lithium, sodium or potassium, leading to the formation of alloys (Sb–Li, Sb–Na or Sb–K) based on the following equations, contributing a high theoretical capacity of 660 mAh g^{-1} in all the LIBs, SIBs and PIBs:

$$Sb + 3Li^+ + 3e^- \rightleftarrows Li_3Sb$$

$$Sb + 3Na^+ + 3e^- \rightleftarrows Na_3Sb$$

$$Sb + 3K^+ + 3e^- \rightleftarrows K_3Sb$$

Although the practical applications of Sb-based anodes have not been realized, the first electrochemical study of Sb-based material happened in the 1970s. The kinetic parameters and thermodynamic abilities of Sb–Li system were originally investigated by Huggins and Weppner with the aid of the electrochemical galvanostatic intermittent titration technique (GITT) and calculations of Gibbs free energy of formation, corresponding enthalpies, and entropies of the alloying system. Based on the observations, Sb showed outstanding chemical diffusion coefficient and the electrochemical reaction could be realized between one mole of Sb and three moles of Li, indicating the promising potential application of Sb anode candidate in LIBs. Unfortunately, the further commercial application of Sb-based anode is still suffering some drawbacks, including the low conductivity and large volume changes of Sb after lithium (147%), sodium (293%), and potassium (400%) insertion. The huge volumetric expansions of Sb-based anodes deriving from the formations of Li_3Sb, Na_3Sb and K_3Sb alloys after fully discharge processes contribute to the generation of the inner stresses within the Sb electrodes, resulting in the pulverization of the Sb bulk materials. The structural damage of the Sb anodes further contributes to the electrical contact loss between the current collector and active electrode materials, leading to the rapid capacity loss and poor cycling life. Similar to other alloy-based electrode materials, the realization of the electrochemical activity for Sb-based anodes also mainly depends on the 'inactive/active' mechanism to mitigate the volume effect of Sb. With continuous development, nanostructured Sb alloys and Sb alloy/carbonaceous material composites appeared as anodes for LIBs in the following years, leading to improved electrochemical performance.

Therefore, the future development of the Sb-based anodes should be the focus on the improvement of the structural stability and enhancement of the conductivity, leading to the overall improved cycling stability and rate performance of the Sb-based anodes in the energy storage systems. Furthermore, advanced fabrication methods with facial synthesis procedures and low-cost are also critically important for the future practical application of Sb-based anode materials.

5.2 Development of Sb anode for Li, Na, K-ion batteries

5.2.1 Sb anode for Li-ion batteries

The study of the electrochemical reactions between Sb alloys and lithium was first started in the 1990s, including Sb–In, Sb–Ti, Sb–Cr, Sb–Fe–Co, and Sb–Co. With the aid of these fundamental research, further improvement and application of Sb-based anodes in LIBs were achieved. In 2004, the Cu_2Sb thin film electrode, which was prepared by Striebel via a pulsed laser deposition technique, showed promising electrochemical performance when used as anode in LIBs [1]. As observed, this Cu_2Sb thin film electrode demonstrated stable cycling stability in the first 50 cycles in the voltage range 0.65–1.4 V vs. Li/Li^+. In 2003, to our best knowledge, the first publication about the carbon modified Sb composite (Sb/C) was reported to be used as the anode in LIBs [2]. After that, from 2004–2010, introducing Sb nanoparticles (NPs) into the porous carbon matrix dominated the research direction of Sb-based anode materials. The introduced porous carbon frameworks not only increase the overall conductivity but also improve the structural stability of Sb NPs. This is mainly due to the volume changes derived from the fully lithiation/delithiation of Sb NPs can be effectively confined within the introduced void spaces, resulting in the improved structural stability of Sb NPs. Additionally, embedding Sb NPs into other different materials with porous structure, high conductivity, or high mechanical strength to improve their overall battery performance will also be discussed in this part. Figure 5.1 shows the relevant summary of the Sb-based anodes for energy storage systems.

5.2.1.1 Sb/C anodes

For the Sb/C composite anode materials, most of the research first focused on mixing Sb NPs with graphite or hard carbon matrix. The graphite or hard carbon can significantly increase the overall conductivity of Sb-based anodes, contributing to the outstanding rate performance [3–17]. For example, the Sb/C nanocomposite and Sb/C/graphite composite were prepared by Hun-Joon Sohn's group via high-energy mechanical milling and were investigated as an anode material for LIBs (figures 5.1(a) and (b)) [18]. According to the x-ray diffraction (XRD) and high-resolution transmission electron microscopy (TEM) results, the Sb/C nanocomposite was composed of nano-sized Sb and amorphous carbon. The electrochemical tests revealed that Sb/C showed outstanding electrochemical reversibility. Additionally, Sb/C/graphite composite exhibited a very high capacity with very good rate performance. A series of nanostructured Sb/C composites have also been successfully synthesized by Limin Wang's group via a simple sol–gel, high-temperature carbon thermal-reduction process [19] (figures 5.1(c)–(l)). For the synthesized Sb/C composites, Sb NPs were uniformly embedded in the pyrolyzed carbon matrix. When used as anodes in LIBs, the best Sb/C composite demonstrated a high initial

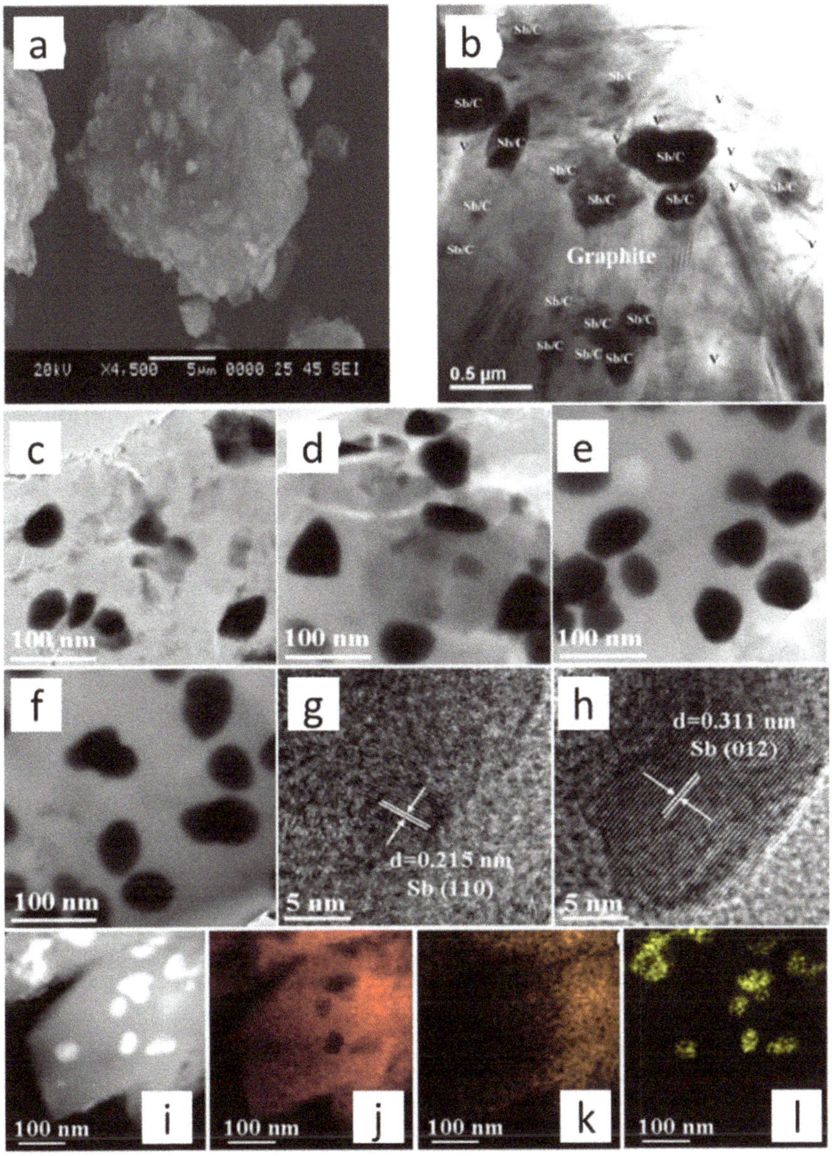

Figure 5.1. (a) SEM and (b) TEM image of Sb/C/graphite composite. Source: [18]. Reproduced with permission of The Electrochemical Society. TEM images of (c)–(f) different Sb/C samples with different Sb contents; (g), (h) high-resolution TEM images; and (i)–(l) EDS mapping of C, O and Sb elements. Source: [19]. Reproduced with permission of Asian Chemical Editorial Society (ACES).

discharge capacity of 1214.6 mAh g^{-1} and a reversible charge capacity of 595.5 mAh g^{-1} with a corresponding initial coulombic efficiency (ICE) of 49% in the first cycle.

However, the long cycling stability and the high ICE of the Sb-based anode cannot be realized by easily mixing Sb NPs and carbon matrix, because the volume changes during charging/discharging processes were not fundamentally addressed by this way. As a result, more and more research paid attention to embed Sb NPs into porous conductive frameworks, such as the porous graphene, reduced graphene oxide (rGO), carbon nanotubes (CNTs) and porous carbon. With the aid of the introduced void spaces, the volume changes of Sb NPs can be well confined within the porous structure, leading to the stable formation of the solid electrolyte interface (SEI) films and improved structural stability. For example, a Sb/graphene (Sb/G) nanohybrid has been synthesized by Shichao Zhang's group through a facile *in situ* solvothermal route employing $SbCl_3$ and graphite oxide as the precursors and $NaBH_4$ as the reducing agent [20]. Microstructural observation showed that the sheet-like nanostructure consisting of Sb NPs attached on few-layer graphene nanosheets (figures 5.2(a) and (b)). The *in situ* attached Sb nanocrystals were uniformly dispersed within the graphene matrix with porous structure and further limited the aggregation of the graphene nanosheets. As observed, the introduction of porous graphene remarkably improved the electrochemical Li storage properties of Sb NPs due to the combined buffering, confining, and conducting effects of graphene. Jim Yang Lee's group reported a facile fabrication of the CNT-encapsulated Sb-based nanorods and this composite contained 7.6 wt% of CNTs and 92.4 wt% of Sb-based nanorods (figures 5.2(c) and (d)) [21]. After the battery test when using this composite as the anode in the LIBs, it showed very high specific capacities and good cycling performance. Chunming Niu's group further fabricated a silica-reinforced carbon nanofibers with encapsulated Sb nanoparticles (denoted as SiO_2/Sb@CNFs) through an electrospinning method (figures 5.2(e)–(m)) [22]. For this composite, the encapsulated Sb NPs within the carbon–silica matrices efficiently buffered the volume changes during Li–Sb alloying–dealloying processes upon cycling and alleviated the mechanical stress. Furthermore, the porous carbon networks contributed fast charge transfer and electrolyte diffusion. Therefore, this Sb-based anode material demonstrated superior lithium storage performance, which showed excellent cycling stability and rate capability, delivering reversible discharge capacities of 700 mAh g^{-1} at 200 mA g^{-1}, 572 mAh g^{-1} at 500 mA g^{-1}, and 468 mAh g^{-1} at 1000 mA g^{-1} each after 400 cycles. With the aid of the *ex situ* as well as *in situ* TEM measurements, it was found that the structural integrity of silica-reinforced Sb-based electrodes could efficiently withstand the mechanical stress induced by the volume changes. Notably, the SiO_2/Sb@CNF//$LiCoO_2$ full cell delivers high reversible capacities of ~400 mAh g^{-1} after 800 cycles at 500 mA g^{-1} and ~336 mAh g^{-1} after 500 cycles at 1000 mA g^{-1}.

Apart from the typical porous structured Sb/C anodes, the yolk-shelled structured Sb/C materials also attracted great research attention. As name suggested, the Sb NP (inner yolk) is encapsulated inside the outer carbon shell (out-shell), and there is a void space between the Sb NP and carbon shell, leading to the typical yolk–shell structured Sb/C anodes. There are several promising advantages for this novel yolk–shell structured Sb/C. Firstly, the void space between the carbon shell and Sb NP can effectively accommodate the volume changes of Sb during lithiation/delithiation processes without damaging the outer carbon shell, leading to the enhanced structural stability of the active Sb electrode materials. Secondly, the introduced carbon shell can effectively increase the overall conductivity of the electrode,

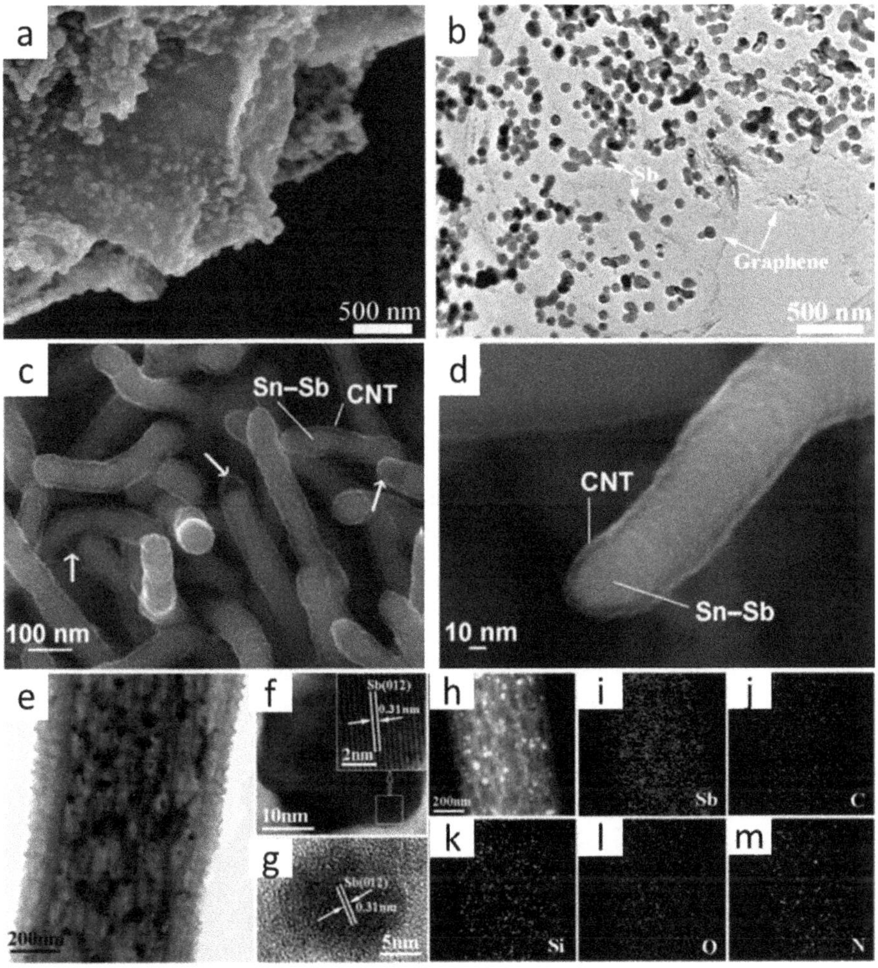

Figure 5.2. (a) SEM images of Sb/G and (b) TEM images of Sb/G, Source: [20]. Reproduced with permission of Elsevier. (c) The thin CNT overlayer of filled nanorods at high magnification, the tips of a few CNTs that have not been completely filled with Sb-based are indicated by arrows, (d) the completely filled tip of a filled CNT. Source: [21]. Reproduced with permission of German Chemical Society (GDCh). (e) TEM image of a single SiO_2/Sb@CNF and (f), (g) HRTEM images of two representative Sb nanoparticles that are embedded within the fiber. (h) HAADFSTEM image of a single fiber with corresponding EDS maps of (i) Sb, (j) C, (k) Si, (l) O, and (m) N elements. Source: [22]. Reproduced with permission of American Chemical Society (ACS).

resulting in the high-rate capacity especially under high current densities. Thirdly, the well confined Sb NP within the carbon shell can limit the formation of the SEI films, leading to the stable formation of the SEI layers and decrease the consumption of the electrolytes, which contributes the high ICE and long cycling life. Therefore, it is an effective way to further develop the yolk-shelled structured Sb/C anodes with long cycling life, high-rate performance, and high ICE.

Zhu's group synthesized the yolk–shell structured Sb@C nanosphere anode for LIBs through a nanoconfined galvanic replacement method (figures 5.3(a)–(i)) [23].

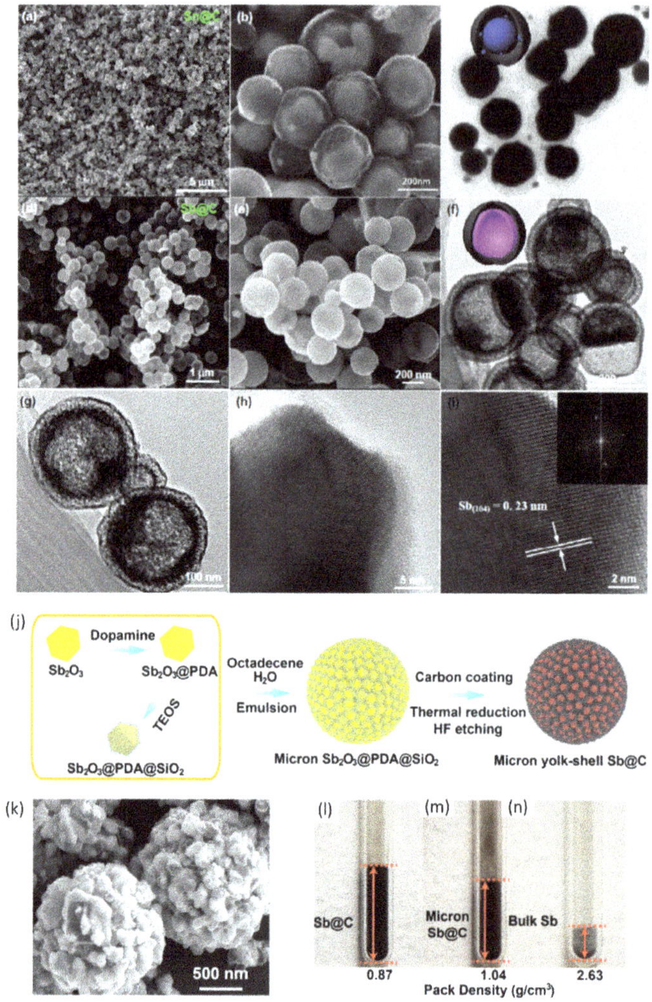

Figure 5.3. SEM and TEM measurements of Sn@C precursor and Sb@C product. (a)–(c) Different-magnification SEM (a), (b) and low-magnification TEM (c) images of Sn@C precursor. (d)–(g) Different-magnification SEM (d), (e) and TEM (f), (g) images of the finally achieved Sb@C product. (h), (i), HRTEM images of Sb nanoparticles comprised of the encapsulated Sb yolks, which clearly display lattice planes of Sb; the inset of panel i shows the corresponding FFT pattern. Source: [23]. Reproduced with permission of American Chemical Society (ACS). Pomegranate Sb@C yolk–shell microspheres (Sb@C microspheres). (j) Schematic illustration of the synthesis of Sb@C microspheres. (k) SEM image of typical Sb@C microspheres. Packing density comparison of (l) the nano-Sb@C yolk–shell structure, (m) Sb@C microspheres and (n) bulk Sb of ~70 micrometers. Source: [24]. Reproduced with permission of The Royal Society of Chemistry (RSC).

The encapsulated Sb hollow yolk was completely protected and confined during the charge/discharge processes by a well conductive carbon thin shell, leading to the improved cycling stability. When used as anode in LIBs, they displayed a large specific capacity (634 mAh g^{-1}), high-rate capability (specific capabilities of 622, 557, 496, 439, and 384 mAh g^{-1} at 100, 200, 500, 1000, and 2000 mA g^{-1}, respectively) and stable cycling performance (a specific capacity of 405 mAh g^{-1} after long 300 cycles at 1000 mA g^{-1}). Another pomegranate Sb/C yolk–shell microspheres were prepared by Xiaolin Li's group and used as a high-performance anode in LIBs with controlled interfacial properties and enhanced packing density (figures 5.3(j)–(n)) [24]. As observed, there was a 78% capacity retention for Sb/C-based LIBs. Moreover, it was also found that the packing density of the microsized yolk–shell structured Sb/C was increased by 30% compared to the nano-sized Sb@C yolk–shell structure. Using these as a material platform, they further elucidated the SEI effect on the cycling stability difference in LIBs. Furthermore, Li's group prepared one-dimensional MoS_2 sheets-decorated $Sb@Sb_2S_3@C(MoS_2\text{-}Sb@Sb_2S_3@C)$ yolk-shelled structure via facile thermal carbonization of Mo–polypyrrole complex coating film [25]. As expected, this MoS_2-$Sb@Sb_2S_3@C$ anode showed very high reversible capacity of 1596 mAh g^{-1} at 0.1 A g^{-1}, high-rate capability of 481 mAh g^{-1} at 5 A g^{-1} and 355 mAh g^{-1} at 8 A g^{-1}, and stable cycling performance of 760 mAh g^{-1} well retained after 100 cycles at 1 A g^{-1}.

5.2.1.2 Other Sb-based anodes

Apart from the Sb/C composite anodes, other Sb-based anode materials also attracted research attention, such as the Sb/TiO_2 [26], Sb/SnO_2 [27, 28], Sb/Sb_2O_3 [29], SiOC/Sb [30], pure Sb electrode (Sb nanocrystals [31], Sb nanochains [32]). Xu Zhang's group successfully synthesized an Sb/TiO_2 composite, in which the outer surface of Sb_2O_3 nanorods were coated by the TiO_2 film with the uniformly embedded Sb particles on the TiO_2 film (figures 5.4(a) and (b)) [26]. This $Sb@TiO_2$ composite demonstrated outstanding rate ability and cycling stability. When served as anode material in LIBs, it contributed a high specific discharge capacity of 609 mA h g^{-1} at a current density of 100 mA g^{-1}. For this composite, the TiO_2 was employed as the outer shell to confine the volume changes of the inside Sb_2O_3 nanorods and Sb nanoparticles. Sb doped SnO_2 (ATO) microspheres composed of ATO nanoparticles were prepared by Ting Chen's group via a hydrothermal process in a nonaqueous and template-free solution from the inorganic precursors ($SnCl_4$ and $Sb(OC_2H_5)_3$) [27]. The physical properties of the as-synthesized samples were investigated by various techniques, including the XRD, SEM, TEM, x-ray photoelectron spectrum (XPS), and N_2 adsorption–desorption isotherms. As observed, the obtained particles had porous structure with highly crystalline ATO microspheres in the diameter range of 3–10 μm (figures 5.4(c) and (d)). The prepared materials showed outstanding cycling life and rate performance when used as the anodes in LIBs. For example, it demonstrated a high initial discharge capacity of 1981 mAh g^{-1} and a charge capacity of 957 mAh g^{-1} within a voltage range between 0.005 V and 3.0 V.

5.2.2 Sb anode for Na-ion batteries

Compared with the deployment of Sb-based anode materials in LIBs, more research attention has been paid to the application of Sb-based anodes for SIBs, due to the better battery performance of Sb-based anode materials in SIBs. Especially since 2013,

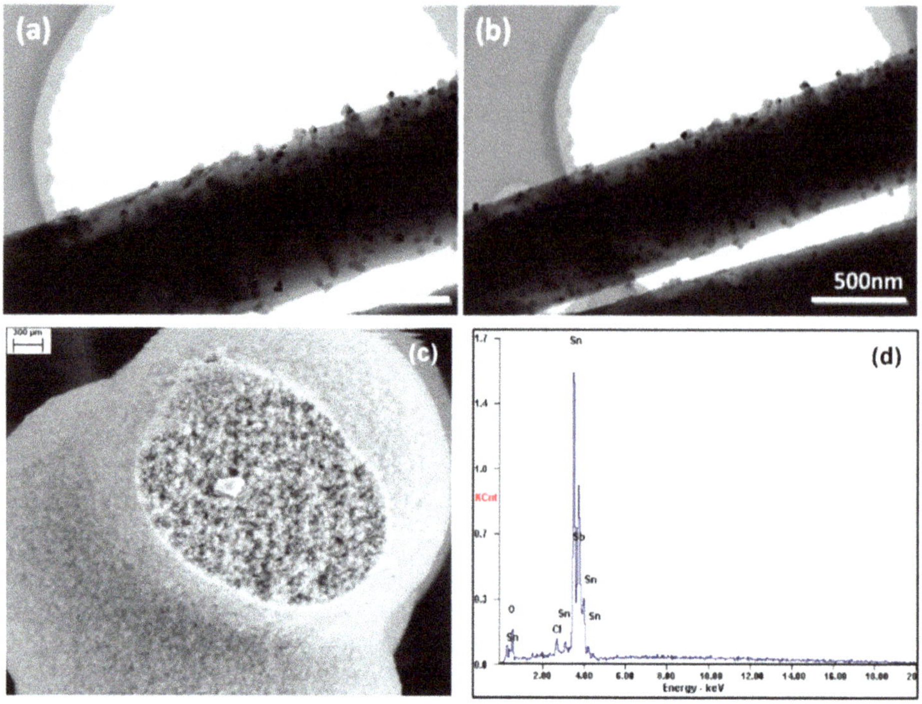

Figure 5.4. TEM images of two different $Sb@TiO_2$ composites (a) and (b). Source: [26]. Reproduced with permission of The Royal Society of Chemistry (RSC). SEM images of ATO microspheres (c) and energy dispersive x-ray spectrum (d). Source: [27]. Reproduced with permission of Elsevier.

the Sb-based electrode materials for SIBs has achieved great progress [33]. Therefore, the detailed development and research about the Sb-based anodes in SIBs will be discussed in this section. Sb materials are highly active of alloying electrodes when used as anodes in SIBs with very high theoretical capacities of 660 $mAh\,g^{-1}$ which corresponding to the reversible formation of Na_3Sb after fully sodiation. Sb-based anodes in SIBs are also suffering from the similar disadvantages which are existing in LIBs, including the large volume changes during charging/discharging processes, continuous SEI formation along with the degradation of the active materials, and poor conductivity. Upon fully sodiation, Sb anodes suffer from large volume extension, leading to the serious pulverization of active Sb particles and less electrical contaction with the copper current collector. Therefore, similar as the improvements of Sb anodes in LIBs, the nanostructure construction, introduction of volume accommodation matrix, and overall conductivity improvement are still the critical strategies to obtain better battery performance of Sb anodes in SIBs. Over the last decade, significant overall electrochemical performance improvements have been achieved for the Sb-based anodes for SIBs. Therefore, it is critical to provide a systematic overview of the material development for Sb-based anodes and contribute valuable proposal to the future deployment of Sb-based anodes for SIBs.

5.2.2.1 Sb/C anodes

Embedding Sb NPs into the porous carbon matrix is an effective way to address the disadvantages of Sb-based anodes. The Sb/nitrogen-doping porous carbon (Sb/NPC) composite using polyaniline nanosheets as the carbon precursor has been successfully prepared by Ji's group (figure 5.5) [34]. The as-prepared Sb/NPC material showed a high reversible specific capacity of 529.6 mA h g^{-1} and a long cycling life with 97.2% capacity well maintained after 100 cycles at 100 mA g^{-1}. A very high-rate capacity of over 350 mA h g^{-1} was retained even at high current density of 1.6 A g^{-1}. The significant improvement of the electrochemical performances of Sb anodes was mainly due to the introduction of the porous NPC framework which provided void spaces to accommodate the volume changes of Sb NPs during cycling and increased the overall conductivity. Another work from Ji's group was about employing the interconnected carbon nanofibers networks (ICNNs) as the carbon matrix to improve the Na storage performance of Sb as anode for SIBs (figure 5.5) [35]. The as-obtained Sb/ICNNs composite demonstrated excellent rate performance and cycling stability. The reversible specific capacity can

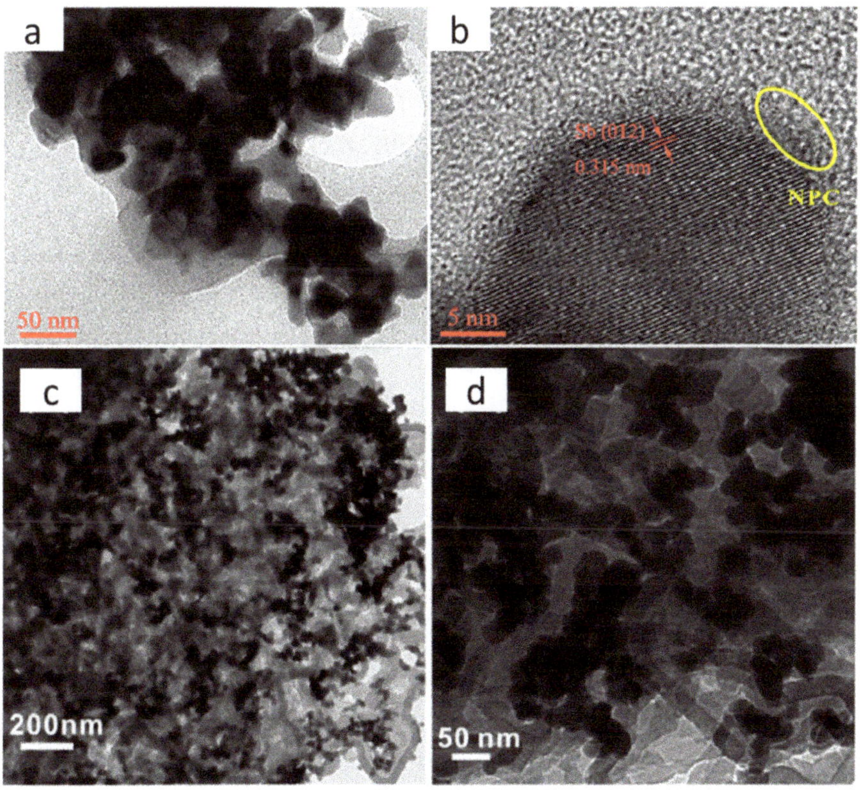

Figure 5.5. TEM and HRTEM images of Sb/NPC composite (a) and (b), Source: [34]. Reproduced with permission of American Chemical Society (ACS). TEM images of the Sb/ICNNs composite. Source: [35]. Reproduced with permission of Elsevier.

be well maintained at about 540 mAh g^{-1} with a very high reversible capacity retention rate of 96.7% after 100 cycles at a current density of 100 mA g^{-1}. Furthermore, the outstanding rate ability was also testified for this composite with a high reversible capacity of 325 mAh g^{-1} at a high current density of 3.2 A g^{-1}. The enhanced cell performances demonstrated that the introduced conductive porous carbon frameworks significantly improved the overall conductivity and structural stability of the Sb NPs, and the proposed idea showed potential to be used as the effective method for other active sodium storage electrodes that are suffering from similar drawbacks.

Apart from the introduction of the porous carbon matrix into the Sb-based electrode systems, coating Sb NPs with carbon layers is also another effective way to improve the battery performance. Different embedding Sb NPs into the porous carbon matrix, generating carbon coating layers directly on the outer surface of Sb NPs can enhance the conductive contact between carbon layers and Sb NPs more effectively, leading to the superior rate ability of the Sb anodes. Xingdong Zhang's group prepared a new Sb-based anode by directly coating polyaniline (PANI) on the surface of Sb NPs anode (Sb-PANI) via an *in situ* oxidative polymerization treatment (figures 5.6(a) and (b)) [36]. When used as anode in SIBs, Sb-PANI showed improved battery performance with an high initial capacity of 510 mAh g^{-1}.

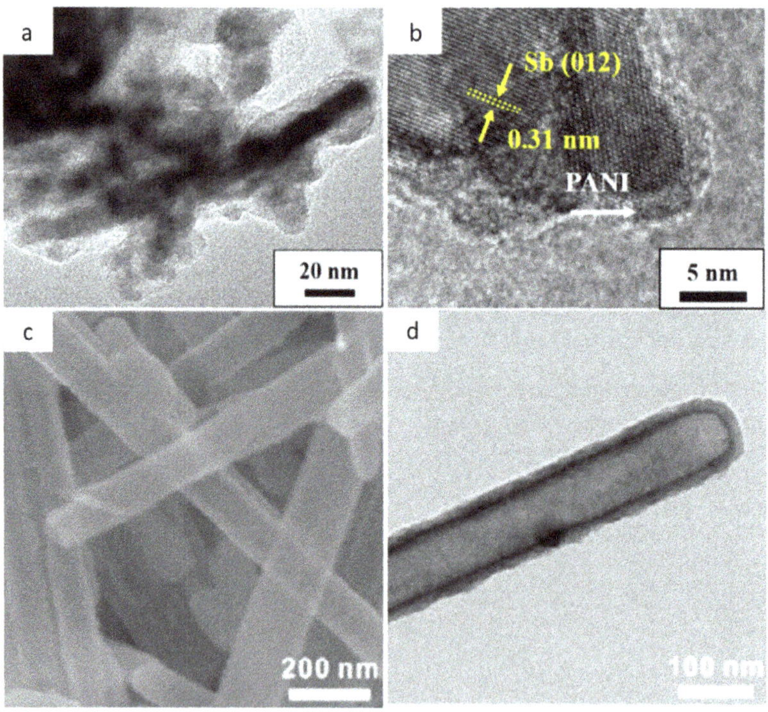

Figure 5.6. TEM (a) and HRTEM (b) images of the Sb/PANI composite. Source: [36]. Reproduced with permission of Elsevier. The SEM (c) and TEM (d) images of the Sb@carbon coaxial nanotubes. Source: [37]. Reproduced with permission of The Royal Society of Chemistry (RSC).

The improved cell performance was mainly due to the synergetic effect between Sb NPs and the uniform conductive PANI coating, which generated rich conductive contact and strengthen protection of the active Sb NPs. Another work reported by Ungyu Paik'g group was also about the carbon coating for the Sb-based anodes via a thermal-reduction strategy. However, different from Zhang's work, Paik's group employed the Sb nanotube as the active material and introduced carbon coating on the outer surface of the Sb nanotube to generate the carbon coated Sb nanotube composite (Sb@carbon coaxial nanotubes, figures 5.6(c) and (d)) [37]. Due to the unique structure, the prepared anode composites demonstrated superior electrochemical sodium storage properties, such as the high irreversible capacity, superior rate ability and stable cycling life, which enabled this new anode material promising to serve as high performance anode candidate for the advanced SIBs.

As we all know, the simple carbon coating on the active electrode materials with large volume changes cannot originally address their poor structural stability, because the outer carbon coating will be damaged after the volume expansion, leading to the continuous formation of the SEI films during reversible cycling. Therefore, similar as the Sb-based for LIBs, the yolk–shell structured Sb/C composites are the most promising design to further improve the battery performance of the Sb-based SIBs. Xiaolin Li's group employed the Sn NP as the active sodium storage material to demonstrate the promising electrochemical advantages of the yolk–shell structure for high performance SIBs (figures 5.7(a) and (b)) [38]. The yolk–shell structured Sb/C was prepared via two steps, including the controlled reduction and selective removal of Sb_2O_3 from carbon coated Sb_2O_3 NPs. The well controlled interspace between the inner Sb NP yolk and outer hollow carbon shell can accommodate the volume expansion of the inner Sb NP after fully sodiation, leading to the improved structural stability and electronic conductivity. As observed, this yolk-shelled Sb/C contributed a high specific capacity of ~ 554 mAh g^{-1} with good rate capability (315 mhA g^{-1} at 10 C rate) and long cycling life (92% capacity retention over 200 cycles). By using the O3-$Na_{0.9}$[Cu0.22Fe0.30$Mn_{0.48}$]O_2 as the cathode to build the full cells found that the full cell delivered a high specific energy of over 130 Wh kg^{-1}, which is around 1.5 times higher than similar design of the full SIB using hard carbon anodes. Furthermore, apart from the typical yolk–shell structures with particle morphologies, one-dimensional (1D) yolk-shelled Si/C composites were also prepared to further increase the conductive contact between the Si-yolk and outer C-shell. Tao's group proposed a new 1D structured yolk-shelled Si/C with Sb NPs well confined within the bamboo-like hierarchical porous N-doped aligned carbon nanotubes (HPACNTs, Sb/HPACNTs) arrays via the combination of solution impregnation, melting diffusion and reduction processes [39]. Due to the introduction of the 1D carbon tubes, this Sb/HPACNTs composite exhibited superior rate performance under high current densities. Furthermore, the well-designed void space within the 1D hollow tubes provided rich space to accommodate the large volume changes of Sb NPs during cycling, resulting in the improved structural stability and long cycling life. Moreover, with the aid of the well dispersed 1D carbon tubes, the aggregation of Sb NPs during electrode fabrication was significantly suppressed, leading to the uniform dispersion of the active Sb NPs.

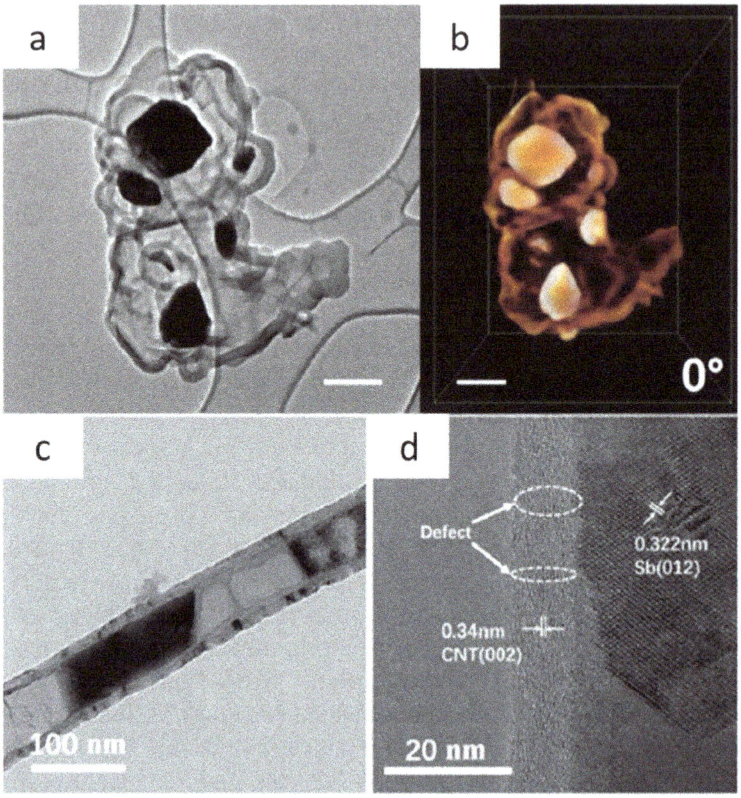

Figure 5.7. (a) TEM image of a typical Sb@C yolk–shell structure, and (b) Screen shots of the 3D reconstruction of the Sb@C structure in (a) showing complete encapsulation of Sb nanoparticles with carbon shells. Source: [38]. Reproduced with permission of Elsevier. The scale bars are 200 nm. (c) TEM image and (d) HRTEM image of the Sb/HPACNTs composite. Source: [39]. Reproduced with permission of The Royal Society of Chemistry (RSC).

Compared with the existing typical sphere-like yolk-shelled Si/C, there are some advantages for the 1D ones, such as the increased overall conductivity, faster ironic/electronic transport, binder free electrode fabrication, and increased tap density.

5.2.2.2 Other Sb/C anodes

Sb can also be combined with other composites to further improve the battery performance of SIBs, such as Ti–O–P [40], Sb_2O_3 [41], CuP_2 [42], NiSb [43], Co [44], Ni [45], $Co(OH)_2$ [46], P [47], SiOC [48], Bi [49]. Jian Yang's group developed 1D yolk–shell Sb@Ti–O–P nanostructures through reducing core–shell Sb_2O_3@TiO_2 nanorods with NaH_2PO_2 (figures 5.8(a)–(c)) [40]. For this Sb@Ti–O–P composite, the volume changes of Sb-yolk were well protected and confined by the outer Ti–O–P shell, leading to the improved structural stability. Additionally, the introduction of the Ti–O–P shell also contributed the formation of the firm SEI films. After the SIBs fabrication, it delivered a high reversible capacity of around 760 mAh g^{-1} at 500 mA g^{-1} after 200 cycles, corresponding to a capacity retention of above 94%.

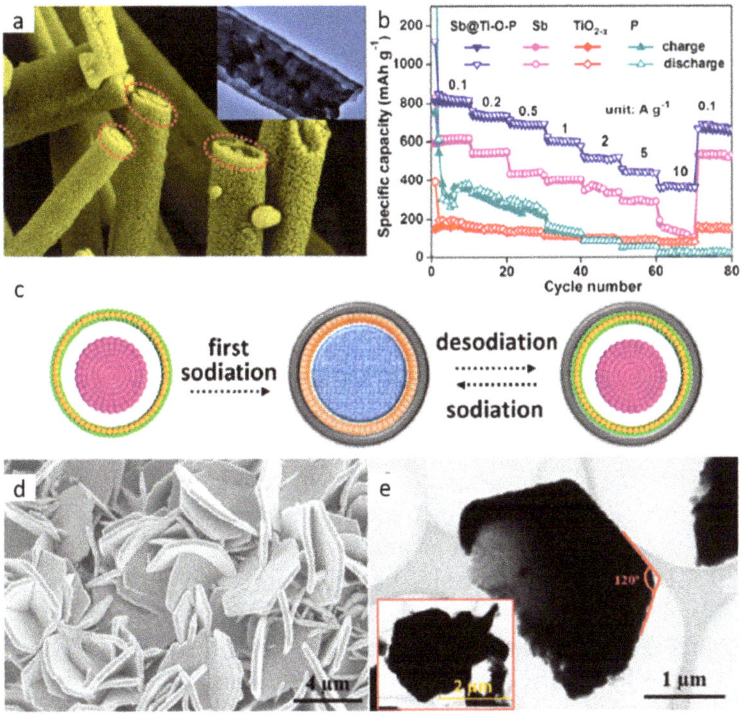

Figure 5.8. SEM (a) with insert TEM images of the Sb@Ti–O–P composite, (b) the rate performance of the Sb@Ti–O–P composite and the reference samples, and (c) schematic illustration on the sodiation/desodiation process of Sb@Ti–O–P. Source: [40]. Reproduced with permission of American Chemical Society (ACS). SEM (d) and TEM (e) images of the Sb@Co(OH)$_2$. Source: [46]. Reproduced with permission of American Chemical Society (ACS).

In particular, it also demonstrated a stable reversible capacity of 360 mAh g^{-1} at a very high current density of 10 A g^{-1}. The author also further fabricated the full SIBs based on this Sb@Ti–O–P composite, which presented a high working potential at about 2.7 V and showed a high capacity of 392 mAh g^{-1} at 1 A g^{-1}, indicating its high electrochemical performance when used as anode in SIBs. Zhonghua Zhang's group reported a core–shell Sb@Co(OH)$_2$ nanosheet anode, which was synthesized through magnetron sputtering Sb NPs onto the mass-productive Co (OH)$_2$ matrix (figures 5.8(d) and (e)) [46]. Therefore, the fabrication method proposed in this work is scalable and available for flow-line production and promising for the future commercial application. For the assembled SIBs, the Sb@Co(OH)$_2$ anode deceived outstanding rate ability of 383.5 mAh g^{-1} at 30 A g^{-1}, high reversible capacity, and increased cycling life.

5.2.3 Sb anode for K-ion batteries

Potassium-ion batteries (PIBs) are also another novel promising energy storage system due to the abundance of potassium resources and their similar electrochemical ability to lithium [50]. Apart from the rich reserve of potassium on the

Earth, the standard reduction potential of potassium (−2.93 V vs E_0) allows PIBs with very high-voltage output, leading to the increased energy density of this system. Unfortunately, compared with LIBs and SIBs, the currently existing popular anode materials are suffering from worse structural stability during K^+ insertion/extraction in PIBs because of the largest ionic radius of K^+. Another critical issue for PIBs is their limited anode candidates due to the large volume changes of the anodes during K^+ insertion/extraction. For example, the currently commercialized graphite anodes show limited reversible capacity and poor cycling life, which seriously limited the further application and scalable production of PIBs.

Among various anode candidates, the Sb or Sb-based electrodes show promising electrochemical performance when used as anodes in PIBs because of their low working potential and high theoretical specific capacity. As mentioned in the beginning of this chapter, the theoretical capacity of Sb anode for potassium storage is 660 mAh g^{-1} along with the generation of K_3Sb alloy after the full K^+ insertion. Compared with the graphite-based anode materials for PIBs, the significant increased specific capacity of Sb can contribute the higher energy density of the full cells. Unfortunately, the Sb-based anodes, which are similar as their application in LIBs and SIBs, have some critical issues when employed in the PIBs system, such as the huge volumetric expansion (around 400%). Therefore, similar approaches, which are suitable for Sb-based anodes for LIBs and SIBs, are still available to be employed for the PIBs to address the volumetric changes induced poor structural stability, including the introduction the porous structure, embedding Sb NPs within the porous matrix, and creating the yolk–shell Sb-based anodes.

5.2.3.1 Porous Sb-based anodes

Directly introducing the porous structure into the pure Sb electrode material is an effective way to address the issue of the volume changes. Yitai Qian's group reported the fabrication of the nano-porous Sb (NP-Sb) through green vacuum-distillation method (figures 5.9(a) and (b)) [51]. The NP-Sb was synthesized through evaporating low-boiling-point zinc (Zn) and the porosity and morphology for the NP-Sb were tuneable by changing the distillation temperature and Zn–Sb composition. Due to the introduction of the nano-porous structure within the active Sb electrodes, the volume expansion was well accommodated, leading to the enhanced structural stability and long cycling life. Additionally, the interconnected frameworks and porous structure also contributed the ionic transfer and electrolyte diffusion, which were helpful for the stable SEI films formation and rate performance.

However, further research found that it does not well address the poor structural stability of the Sb anodes by simply creating void spaces inside the pure Sb particles. Additional carbon coating or embedding Sb NPs within the porous carbon matrix can well address this critical problem, because the introduced carbon frameworks can further strength the mechanical ability of the electrode, resulting in the significantly improved cycling life. Therefore, more research attention has been paid to the design and preparation of the carbon participated porous Sb-based anodes. Chun-Hua Chen's group proposed a novel 3D macroporous Sb@C

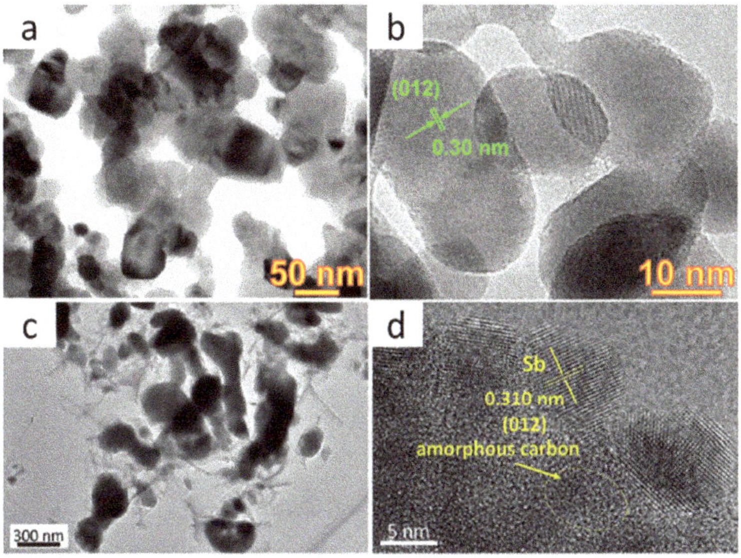

Figure 5.9. TEM (a) and HRTEM (b) images of the NP-Sb (a) and (b). Source: [51]. Reproduced with permission of American Chemical Society (ACS). The TEM (c) and HRTEM (d) image of the Sb@C-3DP anode. Source: [52]. Reproduced with permission of The Royal Society of Chemistry (RSC).

composite (Sb@C-3DP) by a facile KCl template method with a single bi-functional precursor potassium antimony tartrate (figures 5.9(c) and (d)) [52]. As observed, the obtained Sb@C-3DP electrode contributed an outstanding rate ability with 286 mAh g^{-1} at 1 A g^{-1}, and high reversible specific capacity of 516 mAh g^{-1} at 0.05 A g^{-1}. Furthermore, an improved cycling stability was also achieved with 97% capacity retention after 260 cycles, which resulted from the introduced micro-structure that can well accommodate the huge volumetric changes of Sb active particles during cycling processes. More importantly, after the full cell fabrication by coupling the obtained Sb@C-3DP anode with a Prussian blue cathode, very high energy density of 197.6 W h kg^{-1} and outstanding power density of 2067.9 W kg^{-1} were obtained. Liqiang Mai's group reported a new Sb-based anode material with high reversible capacity by *in situ* confining Sb NPs within a 3D carbon framework (3D SbNPs@C) through a template-assisted freeze-drying treatment and subsequent carbothermic reduction [53]. The obtained 3D SbNPs@C composite showed high reversible capacity and improved cycling life when employed as the anode for KIBs. Moreover, the intrinsic mechanism of a K–Sb alloying reaction was also revealed with the aid of the combination of the cyclic voltammetry and *in situ* x-ray diffraction analysis.

Similar to the LIBs and SIBs, the yolk–shell structure design is also suitable to improve the overall electrochemical performance of the Sb-based anodes. Liqiang Mai's group proposed a 1D yolk-shelled Sb/C anode with the segment-like Sb nanorod encapsulated in hollow carbon tube (Sb@HCT, figures 5.10(a) and (b)) [54]. Because of the rich N doping in the carbon tubes, 1D design and the yolk–shell

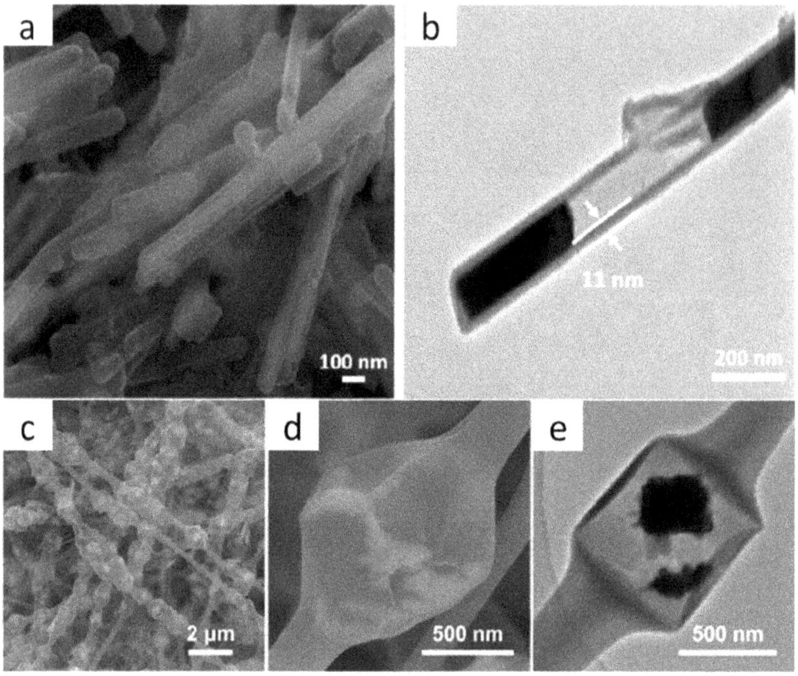

Figure 5.10. SEM (a) and TEM (b) images of the Sb@HCT. Source: [54]. Reproduced with permission of Springer. SEM images (c) and (d) and TEM image (e) of the Sb@CNFs. Source: [55]. Reproduced with permission of Wiley.

advantages, this Sb@HCT anode showed outstanding potassium storage abilities in the case of potassium bis(fluorosulfonyl)imide (KFSI) electrolyte. As observed, this Sb@HCT delivered a high reversible capacity of up to 453.4 mAh g^{-1} at a current density of 0.5 A g^{-1} and superior rate ability with a stable capacity of 211.5 mAh g^{-1} at a very high current density of 5 A g^{-1}. Furthermore, Sb@HCT demonstrates excellent long-cycle stability at a current density of 2 Ag^{-1} over 120 cycles. Meanwhile, it also found that the electrolyte optimization also contributed to the improvement of the electrochemical performance for the Sb-based anode in PIBs. Min Zhu's group also created the Sb-based composite material with hybrid structure of yolk–shell Sb@C nanoboxes embedded in carbon nanofibers (Sb@CNFs) through MOFs-engaged conversion strategy (figures 5.10(c) and (d)) [55]. With the aid of the *in situ* TEM technique, the structural advantages of this new design were fully revealed, and the further structural optimizations were also conducted. As observed, the inner Sb NPs-yolks undergo a significant volume changes during the K^+ insertion/extraction processes. However, thanks to the well-designed void space between the inner Sb-yolks and outer carbon shell, the volume changes of the Sb NPs were well confined within the hollow carbon shell without damaging the outer generated SEI films, leading to the outstanding cycling stability. Additionally, all the yolk-shelled Sb/C units were connected by the carbon fibers, resulting in the fast ironic and electronic transfer within the electrode.

5.2.3.1.1 Challenge of Sb anode for Li, Na, K-ion batteries

Due to advances in structural improvement and composition optimization, significant progress has been achieved in the research of the Sb-based anode materials for the energy storage systems. We summarize the most commonly used Sb-based anode materials, and further conclude their applications in three different energy storage systems, including the lithium-ion batteries, sodium-ion batteries and potassium-ion batteries. We found that the electrochemical performance of the Sb-based anodes can be improved by the three different approaches. Firstly, by introducing the porous structure into the pure Sb electrode can suppress the volume changes of the Sb particles during charging/discharging processes. However, the overall battery performance for the porous pure Sb anodes cannot be fundamentally addressed because the poor mechanical strength of the Sb interconnected frameworks unable to stand the continuous huge volume changes of Sb during cycling. Secondly, designing the porous carbon coated Sb/C composite is another effective way to improve the battery performance of the Sb-based anodes. Thanks to the carbon coating layers on the outer surface of the Sb electrodes, the mechanical strength and overall conductivity of the composite can be significantly improved, leading to the increased rate ability and cycling life. However, the core–shell structured Sb/C composites also suffer from critical issues, because the thin carbon coatings will be damaged after the volume expansion of the Sb particles, leading to the continuous formation of the SEI films and consumption of the electrolyte during cycling. Thirdly, creating the yolk–shell structured Sb-based composites in which the Sb NPs were well encapsulated within the outer hollow shells with a well-designed void space between them. Compared with the pure porous Sb anodes and the porous Sb/C composites, the yolk–shell structured Sb-based anodes showed the best electrochemical performance with outstanding cycling stability and high ICE.

Although great progress has been achieved for the Sb-based anodes in the energy storage systems, there is still a need to develop new techniques and fabrication methods to scalably produce the Sb-based anodes with low-cost and advanced structure designs. For instance, developing the templating free methods to prepare the yolk-shelled Sb-based anodes, and employing low-cost and environmentally friendly chemicals during the whole production processes. In the future, the critical challenges in the development of Sb-based anodes will mainly focus on the following aspects:

(1) The development of the advanced approaches is important to further promote the scalable and commercial application of the Sb-based anodes for energy storage systems. As for the advanced approaches, we believe that three key points should be highlighted, including the environmentally friendly nature, low-cost, and facile treatment processes.

(2) Compared with LIBs and PIBs, the application of Sb-based anodes in SIBs showed more promising electrochemical performance. Therefore, more research attention should be paid to the application of the Sb-based anodes in SIBs, especially for the future commercial SIBs.

(3) Among all the potential Sb-based anode materials, the yolk–shell structured Sb/C composites showed the best cell performance. Therefore, developing new methods to optimize the yolk-shelled Sb/C anodes is a promising way to realize the future commercialization of the Sb-based anodes.

(4) Based on the existing study of the yolk–shell structured Sb/C anodes, some critical disadvantages still need to be addressed, including the poor volumetric energy density because of the introduction of the void space, the limited conductive improvement due to the point-to-point conductive contact between the inner Sb-yolk and outer C-shell, and the complex fabrication methods that usually employ the templating methods to obtain the final structure.

References

[1] Song S W, Reade R P, Cairns E J, Vaughey J T, Thackeray M M and Striebel K A 2004 *J. Electrochem. Soc.* **151** A1012–1019

[2] Dailly A, Ghanbaja J, Willmann P and Billaud D 2003 *Electrochim. Acta* **48** 977–84

[3] Zhang L, Wang Y H, Kan G W, Zhang Z L, Wang C G, Zhong Z Y and Su F B 2014 *RSC Adv.* **4** 43114–20

[4] Guo H P, Ruan B Y, Liu L L, Zhang L, Tao Z L, Chou S L, Wang J Z and Liu H K 2017 *Small* **13** 1700920

[5] Wang C R, Zhang L, Al-Mamun M, Dou Y H, Liu P R, Su D W, Wang G X, Zhang S Q, Wang D and Zhao H J 2019 *Adv. Energy Mater.* **9** 1900909

[6] Zhang L *et al* 2017 *J. Mater. Chem.* A **5** 12073–79

[7] Zhang B W *et al* 2019 *J. Mater. Chem.* A **7** 5214–20

[8] Wang Y, Fan S J, Wu S X, Wang C, Huang Z G and Zhang L 2018 *ACS Appl. Mater. Inter.* **10** 42372–79

[9] Liu X X, Zhang L, Lan X W and Hu X L 2018 *Electrochim. Acta* **274** 23–30

[10] Gui Y H, Liu X, Dou Y H, Zhang L, Al-Mamun M, Jiang L X, Yin H J, He C T and Zhao H J 2019 *Nano Energy* **57** 371–78

[11] Guan P *et al* 2020 *J. Colloid Interf. Sci.* **575** 150–57

[12] Zhang L *et al* 2021 *Energy Environ. Sci.* **14** 3502–9

[13] Zhang L, Dou Y H, Al-Mamun M and Meng G W 2021 *J. Mater. Chem. A.* **9** 19705–9

[14] Wang X C *et al* 2021 *Materials* **14** 2178

[15] Wanshu Luo M C, Pan H, Wang Q, Ma Y, Wang C, Wang Y, Gu Y and Zhang L 2021 *Dalton* T **50** 8330–7

[16] Han N, Li J J, Wang X C, Zhang C L, Liu G, Li X H, Qu J, Peng Z, Zhu X Y and Zhang L 2021 *Nanomater.-Basel* **11** 699

[17] Cao M, Liao F, Wang Q Q, Luo W S, Ma Y L, Zheng X S, Wang Y and Zhang L 2021 *J. Alloy Compd.* **859** 157867

[18] Park C M, Yoon S, Lee S I, Kim J H, Jung J H and Sohn H J 2007 *J. Electrochem. Soc.* **154** A917-A920

[19] Cheng Y, Yi Z, Wang C L, Wang L D, Wu Y M and Wang L M 2016 *Chem. Asian J.* **11** 2173–80

[20] Zhang Y D, Xie J, Zhu T J, Cao G S, Zhao X B and Zhang S C 2014 *J. Power Sources* **247** 204–12

[21] Wang Y and Lee J Y 2006 *Angew Chem. Int. Ed.* **45** 7039–42

[22] Wang H K, Yang X M, Wu Q Z, Zhang Q B, Chen H X, Jing H M, Wang J K, Mi S B, Rogach A L and Niu C M 2018 *ACS Nano* **12** 3406–16

[23] Liu J *et al* 2017 *Nano Lett.* **17** 2034–42

[24] Song J H *et al* 2019 *Nanoscale* **11** 348–55

[25] Li C R, Song H, Mao C M, Peng H R and Li G C 2019 *J. Alloy Compd.* **786** 169–76
[26] Han Q G, Sheng Y L, Han Z W, Li X, Zhang W Q, Li Y and Zhang X 2020 *New J. Chem.* **44** 13430–38
[27] Wang Y and Chen T 2009 *Electrochim. Acta* **54** 3510–15
[28] Wang Y D, Djerdj I, Smarsly B and Antonietti M 2009 *Chem. Mater.* **21** 3202–9
[29] Bryngelsson H, Eskhult J, Edstrom K and Nyholm L 2007 *Electrochim. Acta* **53** 1062–73
[30] Dubey R J C, Sasikumar P V W, Cerboni N, Aebli M, Krumeich F, Blugan G, Kravchyk K V, Graule T and Kovalenko M V 2020 *Nanoscale* **12** 13540–47
[31] He M, Kraychyk K, Walter M and Kovalenko M V 2014 *Nano Lett.* **14** 1255–62
[32] Rodriguez J R, Hamann H J, Mitchell G M, Ortalan V, Pol V G and Ramachandran P V 2019 *ACS Appl. Nano Mater.* **2** 5351–55
[33] Li X Y, Ni J F, Savilov S V and Li L 2018 *Chem.-Eur. J.* **24** 13719–27
[34] Wu T J, Hou H S, Zhang C Y, Ge P, Huang Z D, Jing M J, Qiu X Q and Ji X B 2017 *ACS Appl. Mater. Inter.* **9** 26118–25
[35] Hou H S, Jing M J, Yang Y C, Zhang Y, Song W X, Yang X M, Chen J, Chen Q Y and Ji X B 2015 *J. Power Sources* **284** 227–35
[36] Li M N, Liu Y Y, Qin B, Lu C R, Butt H A, Zheng T and Zhang D X 2021 *J. Alloy Compd.* **861** 158647
[37] Liu Z M, Yu X Y, Lou X W D and Paik U 2016 *Energy Environ. Sci.* **9** 2314–18
[38] Song J H *et al* 2017 *Nano Energy* **40** 504–11
[39] Kang C, Zhou S P, Liu J L, Zhu Y F, Hu A P, Tang Q L, Zhang Y, Chen X H and Tao T 2021 *J. Mater. Chem.* A **9** 2152–60
[40] Wang N N, Bai Z C, Qian Y T and Yang J 2017 *ACS Appl. Mater. Inter.* **9** 447–54
[41] Li N, Liao S, Sun Y, Song H W and Wang C X 2015 *J. Mater. Chem.* A **3** 5820–28
[42] Mun Y S, Yoon Y, Hur J, Park M S, Bae J, Kim J H, Yoon Y S, Yoo I S, Lee S G and Kim I T 2017 *J. Power Sources* **362** 115–22
[43] Zheng X M *et al* 2020 *Nano Energy* **77** 105123
[44] Zhang Y F, Li M, Huang F B, Li Y S, Xu Y Q, Wang F, Yao Q R, Zhou H Y and Deng J Q 2020 *Appl. Surf. Sci.* **499** 143907
[45] Liu J, Yang Z Z, Wang J Q, Gu L, Maier J and Yu Y 2015 *Nano Energy* **16** 389–98
[46] Zhang Y, Gao H, Niu J Z, Ma W S, Shi Y J, Song M J, Peng Z Q and Zhang Z H 2018 *ACS Nano* **12** 11678–88
[47] Zheng X T, Chen K T, Hsieh Y Y and Tuan H Y 2020 *ACS Sustain. Chem. Eng.* **8** 18535–44
[48] Lee Y, Lee K Y and Choi W 2017 *Adv. Funct. Mater.* **27** 1605225
[49] Zhang W, Yan W, Jiang H Q, Wang C, Zhou Y, Ke F H, Cong H J and Deng H X 2020 *ACS Sustain. Chem. Eng.* **8** 335–42
[50] Zhang L *et al* 2019 *Nano Energy* **60** 432–39
[51] An Y L, Tian Y, Ci L J, Xiong S L, Feng J K and Qian Y T 2018 *ACS Nano* **12** 12932–40
[52] Xiao Dong He Z H L, Ying Liao J, Ding X, Hu Q, Na Xiao L and Chen S W A C H 2019 *J. Mater. Chem.* A **7** 9629
[53] Chunhua Han K H, Wang X, Wang C, Li Q, Meng J, Xu X, He Q, Luo W, Wu L and Mai L 2018 *Nanoscale* **10** 6820
[54] Luo W, Li F, Zhang W R, Han K, Gaumet J J, Schaefer H E and Mai L Q 2019 *Nano Res.* **12** 1025–31
[55] Huang H W, Wang J W, Yang X F, Hu R Z, Liu J L, Zhang L and Zhu M 2020 *Angew Chem. Int. Ed.* **59** 14504–10

IOP Publishing

Yurong Ren and Jianmin Ma

Chapter 6

Bismuth as the negative electrode for various batteries

Dan Li

Recently, alkali-ion batteries have attracted extensive attention with the continuous consumption of energy and the worsening of global environment. Among the anode materials, the bismuth-based materials are considered to be attractive due to their higher theoretical capacity with the alloying or conversion/alloying mechanism and large layer spacing along the *C*-axis of bismuth. However, the intractable problem that needs to be solved urgently is that the huge volume change occurrence during the reaction process, which leads to the structural collapse of the material and the degradation in electrochemical performance. In addition, the sluggish ion diffusion kinetics in the charge/discharge process results in poor cycling performance. Therefore, the structural design and composition engineering play a vital role in achieving the high-performance of the bismuth-based materials. In this sense, a review on bismuth-based anode materials for advanced alkali-ion batteries is provided, covering the structural features, ion storage properties, reaction mechanisms, theory analyses, electrochemical performances, and suitable binders and electrolytes.

6.1 Bi anode for various batteries

6.1.1 Bi properties

The theoretical specific capacity of bismuth is lower than that of tin and antimony (antimony: 660 mA h g^{-1}; bismuth: 385 mA h g^{-1}; tin: 847 mA h g^{-1}). At the same time, a large volume expansion of more than 400% is generated because of the embedding of ions into the lattice. In terms of working voltage, although the discharge platforms of alloying reaction of bismuth are relatively high, the generation of dendrite could therefore be inhibited, which ensures the safety of the battery within cycling. Furthermore, bismuth has strengths of low price, environmentally friendly, non-toxic, and high volumetric capacity (3765 mA h cm^{-3}). More importantly, its unique lamellar crystal structure, i.e., the large layer spacing along the *C*-axis

doi:10.1088/978-0-7503-4871-3ch6

(d(003) = 3.95 Å), is conducive to the insertion and extraction of ions. These characteristics make bismuth-based materials, including metallic Bi, and its chalcogenides, alloys as well as other compounds, the potential anode materials for rechargeable ion batteries.

At present, the synthetic approaches for preparing bismuth-based anode materials, include the sol–gel method, hydrothermal method, solvothermal method, electrospinning method, thermal reduction method, and so on. Among them, the thermal reduction method is a common and easy to operate way to fabricate bismuth-based materials. Electrospinning is usually applied for the preparation of one-dimensional bismuth-based nanofiber structured composites by electrostatic force. The resulting composites are composed of carbon and bismuth-based materials and usually have porous or hollow structures. The solvothermal method is also an effective method to synthesize bismuth-based materials with small size distribution, which is facile to obtain bismuth-based nanomaterials with high crystallinity and purity.

Like other alloy materials, the serious problem of bismuth-based materials is the inevitably produced large volume changes during charging and discharging process, which leads to the pulverization of electrode material in the electrochemical reaction process, and gradually loosing the contact from the current collector, and thus a large capacity attenuation. To address this dreadful issue, the structural stability and electrochemical performance of bismuth-based materials can be improved by the following three methods:

(1) Reducing particle size. The size of the particles has a direct impact on the crystal packing, resulting in steric effects and thus affecting its electrochemical properties. Small particle size (in nanometer scale) can not only shorten the ion diffusion distance, reduce the particle surface stress, but also increase the active sites, increasing the contact area between the material and the electrolyte. While the materials with large grain size are prone to collapse, fracture, and pulverization, which lead to the consequently capacity fading during the reaction process. Notably, the merely nanocrystallization of particle size is not effective enough to maintain the stability of structure and properties based on the previously studies. In addition, the high specific surface of nanostructured materials will also lead to the low charge–discharge efficiency in the first several cycles, and excessive consumption of the limited ions in the battery, resulting in a decrease in capacity.

(2) Compositing with carbon-based materials. The composite with carbon material can effectively restrain the volume expansion of the bismuth-based materials during the discharging process and improve the electrical conductivity of the composite. The common composite forms of hybridizing are mixing with carbon-based materials, encapsulation in carbon shells, stacking within two-dimensional (2D) layered carbon-based material, encapsulation in one-dimensional (1D) rod-like carbon-based material, encapsulation in a three-dimensional (3D) carbon matrix material. However, it is necessary to optimize the carbon content in the composite. If the carbon content is too low, the insufficient carbon cannot effectively buffer the volume

expansion/extraction of the bismuth-based materials, otherwise, the over-dosed carbon will result in the sacrifice of the theoretical specific capacity of the composite. Previous studies have shown that incomplete or unstable coatings cannot maintain their integrity in a long cycle period. When the cycle increases to a certain number, the coatings will gradually break and lose the coating effect, leading to the pulverization of the electrode material. Therefore, solid, complete, and effective coating modification can maintain the structural stability of the material.

(3) Alloying with other metals. Introducing other metal elements to form the alloys can not only buffer the structural change of bismuth-based materials, but also increase the electrical conductivity of the composite. According to whether the introduced metal has the electrochemical activity of ion storage, it can be divided into electrochemically inert metals and electrochemically active metals. For the electrochemically active metals, each component in the obtained alloy could contribute to the capacity within the reaction. While the electrochemically inert metals only serve as the conductive agent, although it would not alloy with ions within the discharging process. It should be noted that the above three modification methods are usually combined to inhibit the volume change of bismuth-based materials.

In this chapter, we focus on the bismuth-based materials that underwent the conversion and alloying mechanisms and provide an overview of the reports on their ion storage properties, specific reaction mechanism, the suitable and applied binders and electrolytes, and the theory analysis on these bismuth-based anode materials.

6.2 Development of Bi anode for Li, Na, K-ion batteries

6.2.1 Bi anode for Li-ion batteries

In terms of the reaction mechanism, the metallic bismuth alloys three lithium ions that form Li_3Bi by achieving a relatively low theoretical capacity of 385 mA h g^{-1} due to its high atomic mass in group VA. But the distinctive advantage of high volumetric capacity of 3765 mA h cm^{-3}, which is much higher than that of silicon (2190 mA h cm^{-3}), tin (1991 mA h cm^{-3}), as well as antimony (1889 mA h cm^{-3}) [1], make bismuth attractive to be the anode material for lithium-ion batteries. Furthermore, the formation of Li_3Bi occurred at the potential of around 0.8 V makes the reaction is free from the lithium plating. For the Bi_2X_3 compounds, including Bi_2O_3, Bi_2S_3, and Bi_2Se_3, the reaction mechanism composes the conversion and alloying processes. Specifically, the Bi_2X_3 compounds undergo the conversion reaction to generate the metallic bismuth, which will go through the alloying procedure to form Li_3Bi.

6.2.1.1 Metallic bismuth-based composites

Tarascon's group firstly reported the electrochemical properties of nanometric bismuth (about 20 nm) via the direct electrodeposition at room temperature. Considering the relatively large volume expansion of bismuth after alloying reaction, the structure designing played a significant role in the promotion of the electrochemical performance. Hou and Peng confined the bismuth dots into the spongiform carbon network with porous structure via the carbothermal reduction method, as shown in figure 6.1(a) [2]. The obtained Bi@PC composited exhibited a relatively high capacity of 520 mA h g^{-1} at a current density of 0.2 A g^{-1} and good long-term cycling property at 0.5 A g^{-1} with a capacity of 380 after 500 cycles. And

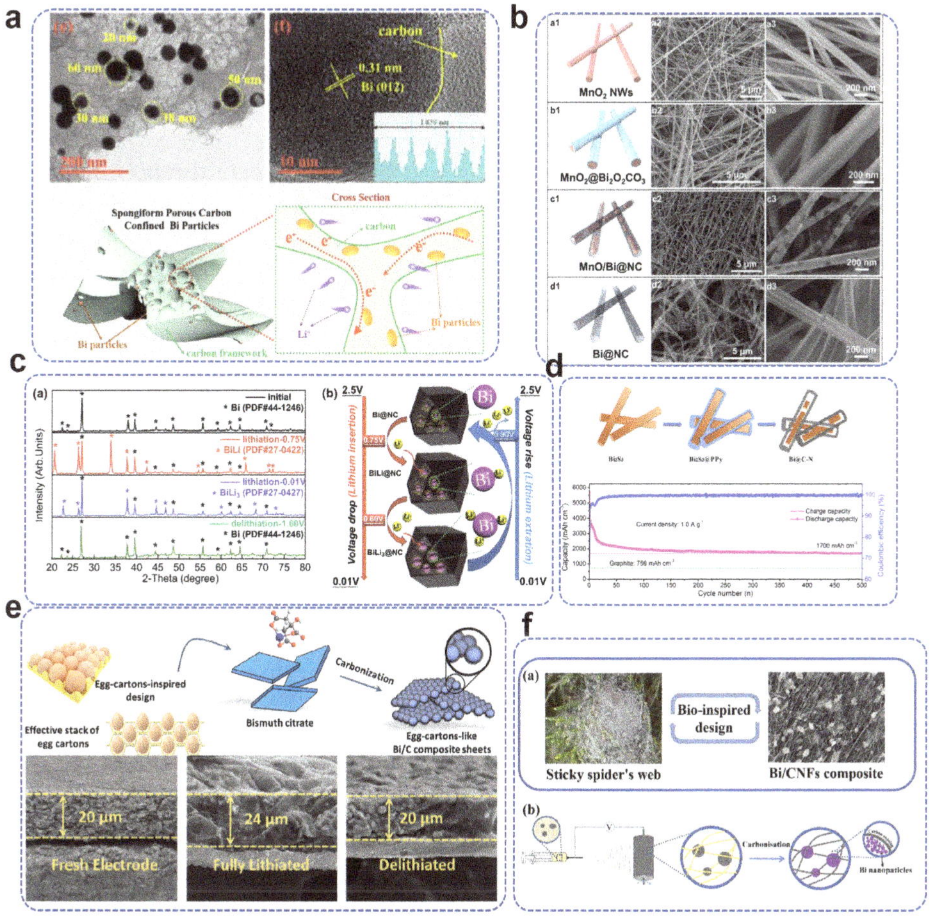

Figure 6.1. (a) TEM images and the schematic illustration of the fabrication of the spongiform porous Bi/C composite constructed by one-step carbothermal reduction method. Source: [2]. Reproduced with permission of Wiley-VCH Verlag. (b) Schematic illustration of synthesis procedure and SEM images for MnO_2, $MnO_2@Bi_2O_2CO_3$, MnO/Bi@NC, and Bi@NC. Source: [3]. Reproduced with permission of Royal Society of Chemistry. (c) XRD patterns and schematic illustration of the chemical evolution of the Bi@NC electrode. Source: [4]. Reproduced with permission of Springer Singapore. (d) Schematic illustration of synthesis procedure and cycling performance for the Bi@C–N. Source: [5]. Reproduced with permission of American Chemical Society. (e) Schematic illustration of the egg-carton-inspired design and the synthesis procedure of Bi/C, as well as the cross-sectional SEM images of the electrode at different states. Source: [6]. Reproduced with permission of American Chemical Society. (f) Schematic illustration of the bio-inspired design and the synthesis procedure of Bi/CNFs. Source: [7]. Reproduced with permission of The Royal Society of Chemistry.

the satisfying performance can be attributed to the 3D porous structure. Firstly, the sponge-like carbon frame increased the connection interface of the electrode/electrolyte to ensure the sufficiently wetting of the electrode. Secondly, the large inner pores in the carbon offered the extra spacing for the volume expansion after the lithium-ion intercalation. Thirdly, the carbon framework fully enclosed the bismuth dots to restrain the bismuth agglomeration and expansion. The same concept of bismuth dots was also applied by Wu's group, in which the bismuth dots were encapsulated into the nitrogen-doped carbon nanotubes to form a composite of Bi@NC [3]. The composite was synthesized via a facile template strategy by using the MnO_2 as the template and the following selective etching treatment. The as-prepared ultralong carbon tubes with a large aspect ratio is beneficial for the fast electron transfer along the continuous longitudinal channels (figure 6.1(b)). Wang's group employed the replacement reaction method to fabricate the bismuth nanoparticles embedded in the nitrogen-doped porous carbon as the anode material for lithium-ion batteries, which could enhance the electrical conductivity and provide the free void space to restrain the aggregation of the nanoparticles and relieve mechanical strain of bismuth upon lithium insertion and extraction process, as displayed in figure 6.1(c) [4]. As a result, the as-prepared material achieved better rate and cycling performance compared with the simple composite with carbon or the bare bismuth materials. Furthermore, the phase evolution was also explored by the *ex situ* XRD measurements, revealing the transformation from metallic bismuth to BiLi and $BiLi_3$ occurred at the potential of 0.75 and 0.6 V within the lithium ion insertion process, respectively.

For the extraction process of lithium ions, the reconversion reaction between $BiLi_3$ to Bi takes place in 0.9 V vs. Li^+/Li. Except the bismuth dots, Hou's group also reported a bismuth@N-doped carbon composite with yolk–shell structure through the *in situ* thermal reduction of polypyrrole that coated on the Bi_2S_3 nanorods, which showed a stable volumetric capacity of 1700 mA h cm^{-3} over 500 cycles at 1.0 A g^{-1}, nearly 2.2 times that of graphite (figure 6.1(d)) [5]. And the good electrochemical properties are owing to the enhanced electrical conductivity stemmed from the nitrogen-doping and the reduced diffusion pathway for lithium ions originated from the 1D structure of bismuth nanorods. Moreover, the void cavity within the carbon shells would supply the free space for the structural swelling during the ions intercalation processes and therefore guarantee the stability in structure and therefore the cycling performance. Inspired by the structural characteristics of egg cartons, Yu *et al* designed a Bi/C composite composed of the compacted 2D carbon frames densely surrounding the bismuth nanoparticles by the pyrolysis approach. And the schematic illustration of the design and preparation procedure of the composite is shown in figure 6.1(e) [6]. The advantage of the egg-carton-inspired design ensured the high tap density and limited side reactions without the excessive surface area. Therefore, the composite delivered a volumetric capacity of 1461 mA h cm^{-3} at 0.1 A g^{-1}, and a high capacity retention of 89% after 1000 cycles at 1 A g^{-1}. The composite presented stable cycling ability, which was disclosed by the cross-sectional SEM observation at the fully lithiation and delithiation states. And the results turned out that change in thickness between these two states was only 20%, which is much smaller than the theoretical value of 215%, meaning that the structure could effectively buffer the volume variation of bismuth. Another good example in structural design was proposed by Yang *et al* with a biomimetic strategy by applying the sticky spider web to achieve the high gravimetric and volumetric energy density (figure 6.1(f)) [7]. The electrospinning technique was adopted to prepare the over 50 wt% bismuth nanoparticle sealed-in free-standing carbon nanofibers. The carbon nanofibers tightly enclosed the bismuth to afford the fast electron diffusion channels, leading to a good relatively high-performance of the composite with a capacity of 483 mA h g^{-1} at 100 mA g^{-1} upon 200 cycles.

6.2.1.2 Bismuth oxides

Owing to the relatively simple synthesis methods, suitable working window, as well as the high theoretical capacity (690 mA h g^{-1}) and volumetric capacity (6280 mA h cm^{-3}) [8, 9], bismuth oxide (Bi_2O_3) has raised concern in the related areas of energy. It is known that Bi_2O_3 has five polymorphic crystal types, namely α-, β-, γ-, δ- and ω-Bi_2O_3, in which the monoclinic α-phase is the most common phase due to its high stability, and the metastable β-Bi_2O_3 possesses the highest ionic conductivity [10].

The first study of the Bi_2O_3 in the field of lithium-ion batteries was conducted by Luo's group [9]. A composite was fabricated through the directly growth of Bi_2O_3 particles on the nickel foam via the polymer-assisted solution method and obtained a free-standing electrode, which presented improved electrochemical properties compared with the commercial Bi_2O_3 and Bi_2O_3 powder. And the better performance of the as-prepared composite could be attributed to the thin carbon shells covered on the particles, which interknitted and attached on the nickel foam surface. Sun *et al* [10] reported a heterostructured 2D Bi_2O_3 (α-/β-phases) nanosheets implanted on the nanotubes and studied its lithium storage properties (figure 6.2(a)). The dual-phase Bi_2O_3 with the heterointerface showed a narrowed band gap to ensure an increased electrical conductivity. And the 2D structure provided more exposed active surface, fast ion transport, as well as the decreased ion/electron pathway. As a result, the heterostructured 2D Bi_2O_3/CNTs reached a capacity of 823.5 mA h g^{-1}, which corresponded to an increase of about 27% compared with the bare Bi_2O_3. Fang *et al* [11] synthesized a composite consisting the Bi_2O_3 nanoparticles encapsulated by 3D porous nitrogen-doped graphene (Bi_2O_3@N-G), and the composite was provided with a relatively large surface area of 112 m^2 g^{-1} from the 3D porous structured carbon network, resulting in an accelerated transportation for ions and electrons and a good rate capability of 273.5 mA h g^{-1} at a relatively high current density of 10 A g^{-1}, as shown in figure 6.2(b). Elumalai's group proposed a composite consisting the carbon quantum dot derived from the spoiled milk and Bi_2O_3 nanoparticles via a facile hydrothermal reaction (CQD–Bi_2O_3) [12]. And the composite presented a high electrochemical activity and delivered a capacity of 1500 mA h g^{-1} at 0.2C rate (figure 6.2(c)). In addition, Yao *et al* fabricated a Bi_2O_3/Bi nanocomposite confined by N-doped honeycomb-like porous carbon (Bi_2O_3/Bi@NHPC) via a chemical blowing method to embed the Bi_2O_3 and Bi nanoparticles [13]. The honeycomb-like porous carbon matrix could protect the nanoparticles from cracking and the electrode material from pulverization and then

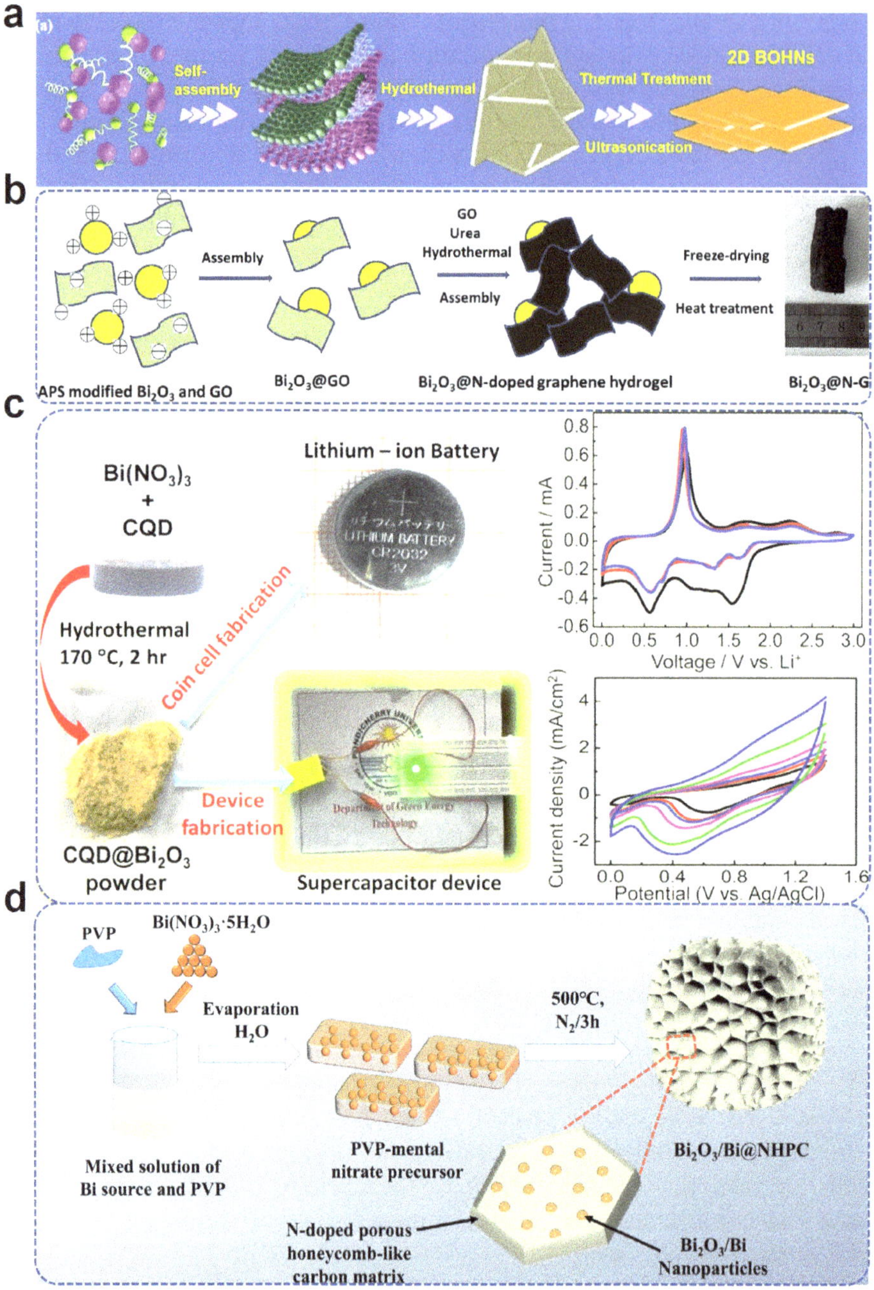

Figure 6.2. (a) Schematic illustration of the fabrication of the 2D Bi_2O_3. Source: [10]. Reproduced with permission of American Chemical Society. (b) Schematic illustration of the fabrication of the Bi_2O_3@N-G. Source: [11]. Reproduced with permission of Elsevier. (c) Schematic illustration of the fabrication and CV curves of the CQD–Bi_2O_3. Source: [12]. Reproduced with permission of American Chemical Society. (d) Schematic illustration of the fabrication of Bi_2O_3/Bi@NHPC. Source: [13]. Reproduced with permission of Elsevier B.V.

peeling off from the current collector considering the large volume variation. Furthermore, this kind of 3D carbon network could offer sufficient ion diffusion channels. As a result, the obtained composite delivered high reversible capacity of 408 mA h g^{-1} upon 1000 cycles at 1 A g^{-1}, and displayed a good rate capability of 258 mA h g^{-1} at 10 A g^{-1} and a decent capacity retention of 78% at 5 A g^{-1} over 2000 cycles.

6.2.1.3 Bismuth sulfides

Bismuth sulfides (Bi_2S_3) are regarded as a type of potential candidates for the lithium-ion anode materials on account of the high volumetric capacity of 6280 mA h cm^{-3}, narrow band gap of 1.3 eV, and the unique laminar structure [14, 15]. Within the discharge process, the lithium-ion intercalation involves two procedures of conversion reaction consuming 6 Li^+ to generate the metallic bismuth with a volume expansion of 90% and the following alloying process taken part in with 3 Li^+, resulting in a further 74% volume increase [16]. The resulting large volumetric volume change upon the lithium insertion and extraction is dreadful for the structural stability of Bi_2S_3 with the consequence of cracking and the final pulverization of the electrode material, leading to the drastic degradation in capacity. Furthermore, the sluggish ion diffusion is also a weakness for the Bi_2S_3. To circumvent these problems, many efforts have been devoted into the structure modification by compositing with carbon-based materials.

For instance, Li's group designed intimately assembled Bi_2S_3 nanowires encapsulated in polypyrrole shells (Bi_2S_3-PPy), which were featured with abundant void space and open channels for lithium-ion diffusion, as shown in figure 6.3(a) [17]. In addition, the flexible and elastic polypyrrole shells took the responsible to tolerance the volume swelling and shrinking to make sure maintaining the stability both in structure and capacity. As a result, the composited exhibited a high-capacity retention of 99% after 500 cycles and good rate capability with a capacity of 337 mA h g^{-1} at 10 C. Zhang and Li *et al* reported a composite of Bi_2S_3 nanorods hosted on reduced graphene oxide nanosheets (Bi_2S_3/rGO) via the pyrolysis of the molecular precursor (figure 6.3(b)) [18]. Owing to the short ion diffusion channel provided by the 1D structure of the nanorod architecture and the closed interaction with the rGO layer, the composited equipped with good kinetic characteristics and thus exhibited a capacity of 558.4 mA h g^{-1} at 0.2 mA g^{-1} after 200 cycles, which was much higher than that of bare Bi_2S_3.

Although rGO layer coating could effectively address the severe capacity degradation by restraining the volume change and accelerating the diffusion transfer, it is hard to coat on the surface of Bi_2S_3 because of the intrinsic hydrophobicity of rGO and the strong p-interaction of rGO nanosheets. Therefore, Song's group used the 2D $(BiO)_2CO_3$ nanosheets as the starting material, which was subjected to a modification by the cationic surfactant of cetyltrimethylammonium bromide to synthesize a sandwich-like graphene–Bi_2S_3 hybrids (rGO@Bi_2S_3) through an electrostatic self-assembly process (figure 6.3(c)) [19]. The close wrapping between Bi_2S_3 and graphene layer presented robust structure stability with plenty of space for the volume expansion/extraction and multidimensional pathway for lithium ions,

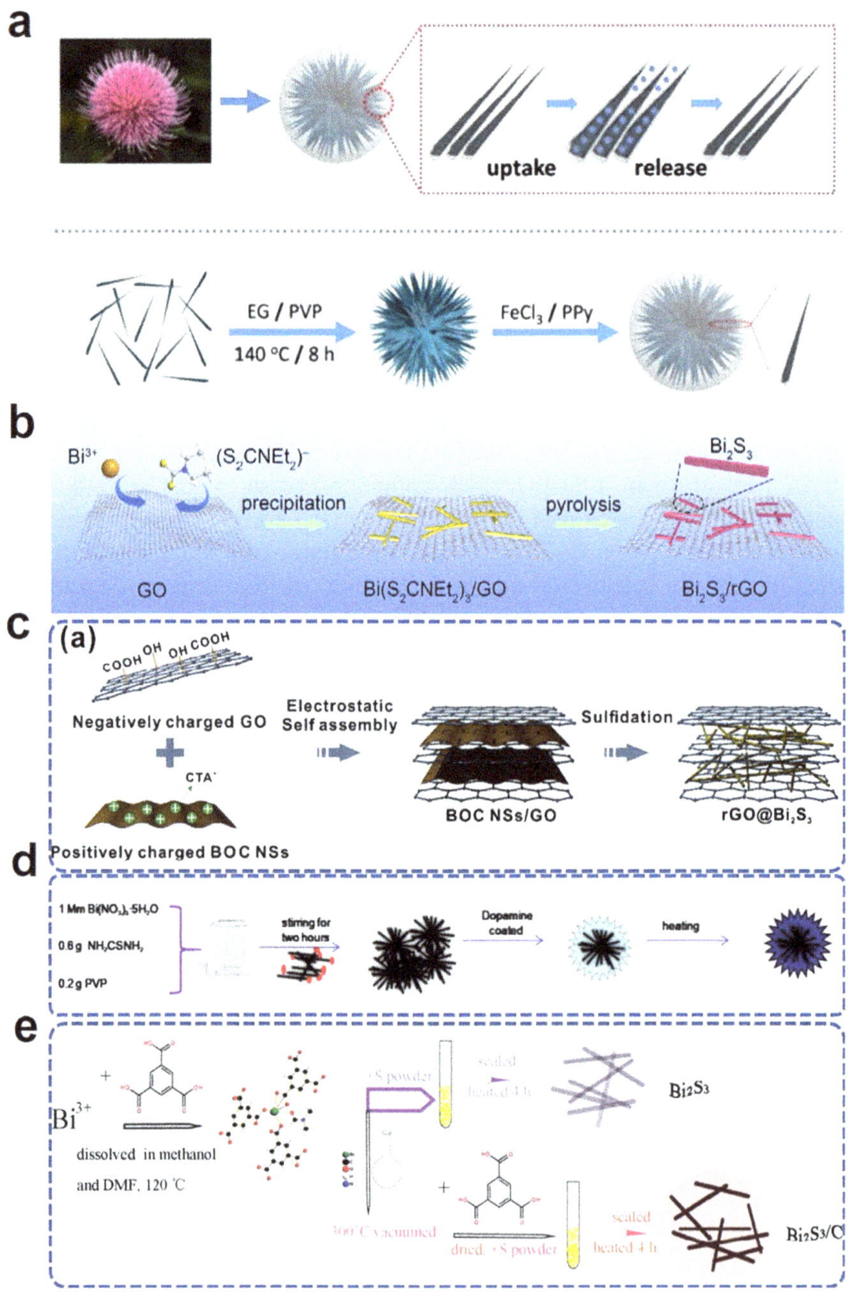

Figure 6.3. Schematic illustration of (a) design principle of the architecture inspired by cirsium and synthesis procedure of Bi_2S_3-PPy. Source: [17]. Reproduced with permission of Elsevier B.V. (b) The fabrication of Bi_2S_3/rGO. Source: [18]. Reproduced with permission of Wiley. (c) The fabrication of rGO@Bi_2S_3. Source: [19]. Reproduced with permission of Elsevier B.V. (d) The fabrication of N–C@Bi_2S_3. Source: [20]. Reproduced with permission of Royal Society of Chemistry. (e) The fabrication of Bi_2S_3 and Bi_2S_3/C. Source: [21]. Reproduced with permission of Royal Society of Chemistry.

and thus good electrochemical properties with a high capacity of 657 mA h g^{-1} at 125 mA g^{-1} after 100 cycles. Peng and Li *et al* adopted the dopamine as the carbon source to cover the carbon layer on the surface of the Bi_2S_3 (N–C@Bi_2S_3) within the polymerization process and obtained a composite of nitrogen-enriched carbon coated flower-like Bi_2S_3 via a hydrothermal approach (figure 6.3(d)) [20]. The N-doped carbon layer derived from dopamine could physically and chemically immobilize the soluble polysulfide species to prevent the composite from the capacity degradation coupling with the merit of increased electrical conductivity and restrained volume change. The composite presented a high reversible capacity of 1001.7 mA h g^{-1} at 200 mA g^{-1} after 224 cycles with a retention in capacity of 78.5%. While Tang *et al* applied the inexpensive and easily handled trimesic acid, as an organic–inorganic material with a framework structure [22], to be a novel carbon source and solvent to synthesize a Bi_2S_3/C composite, which played a part in the suppression of the particle aggregation (figure 6.3(e)). And the obtained composite displayed a good rate capability and high capacity [21].

6.2.1.4 Bismuth selenide

Bismuth selenide (Bi_2Se_3), crystallizing in layered structure, allows lithium ions intercalation within the layers and makes it applicable as the anode material for lithium-ion batteries, considering the higher electrical conductivity compared with its oxide and sulfide counterparts. But the relevant study in the lithium storage is rare, possibly due to the low theoretical capacity of 491 mA h g^{-1} or/and sluggish ion kinetics. Under this premise, the structural and composition design is essential for achieving decent lithium storage performance. For instance, Huang *et al* introduced graphene nanosheets into the Bi_2S_3 nanosheets utilized the vacuum filtration method and a following thermal reduction treatment to construct the conductive framework to alleviate the volume swelling/shrinking and prevent the nanosheets from stacking and aggregating [23]. And the flexible composite delivered a capacity of reversible capacity of 203 mA h g^{-1} under 50 mA g^{-1} after 100 cycles. In addition, Lu *et al* reported a CNTs@C@Bi_2Se_3 nanocomposite using CNTs to band Bi_2S_3 nanosheets to build the conductive electron transfer network [24]. And the amorphous carbon coating connected the CNTs and Bi_2S_3 nanosheets intimately to buffer the volume change within the ion insertion and extraction, leading to a high capacity 432 mA h g^{-1} under 100 mA g^{-1} after 100 cycles and a good rate capability.

6.2.1.5 Other bismuth-based materials

There are several other bismuth-based materials applied as the anode materials, such as Bi_2MoO_6, Bi_2WO_6, $Bi_5Nb_3O_{15}$, $BiVO_4$ and so on. The multicomponent-metal oxides have gathered increased interest in the energy storage field due to their relatively high electrical conductivity and good properties owing to the instinct synergistic effect between different components [30]. Among them, the Bi_2MoO_6, with photoelectric characteristics composed of alternate $[Bi_2O_2]^{2+}$ layers and $[MoO_4]^{2-}$ perovskite layers, is appealing to store lithium ions with low equilibrium potential for reaction below 1.0 V and high theoretical capacity of 791 mA h g^{-1}. Xu *et al* composited the positive-charged Bi_2MoO_6 with a MXene of negative-charged

$Ti_3C_2T_x$ to improve the structural stability and electrical conductivity due to the low Li^+ diffusion barrier of nearly 0.07 eV [31]. The $Bi_2MoO_6/Ti_3C_2T_x$, prepared by the electrostatic self-assembly and freeze-drying methods, possessed a plate-to-layer heterostructure, which was beneficial to increase exposed active sites and well-restrained volume change due to sufficient interconnected Bi_2MoO_6 and $Ti_3C_2T_x$ by avoiding the aggregation and restacking of both Bi_2MoO_6 and the MXene, as shown in figure 6.4(a) [25]. As a result, the heterostructured material exhibited enhanced rate performance and cycling durability, showing a capacity of 328.2 mA h g^{-1} under 2 A g^{-1} and 545.1 mA h g^{-1} and 99.6% Coulombic efficiency at 1 A g^{-1}. Moreover, Liu's group reported a composite consisting Bi_2MoO_6 encapsulated by reduced graphene oxide by the one-pot hydrothermal method. The composite was provided with synergistic effect between Bi_2MoO_6 and rGO nanosheets (Bi_2MoO_6/rGO), enhanced electrical conductivity and structural durability, leading to a much higher capacity retention of 80.4% compared to that of bare Bi_2MoO_6 (figure 6.4(b)) [26]. Guo *et al* explained the decent lithium-ion storage of Bi_2MoO_6 from the view of crystalline by the *in situ* synchrotron x-ray powder diffraction measurements and the density functional theory (DFT) calculations, revealing that the as-synthesized atomically thin Bi_2MoO_6 was instinctly good at the rate capability due to the interlayer coupling effect and 2D configuration. Furthermore, the unbalanced charge distribution between $[Bi_2O_2]^{2+}$ and $[MoO_4]^{2-}$ layers brought in the built-in electric fields to accelerate the ion/electron diffusion, as shown in figure 6.4(c) [27].

In terms of $Bi_5Nb_3O_{15}$, Shu *et al* reported a composite of $Bi_5Nb_3O_{15}$@C with the particle size of 0.5–2.0 μm using simple glucose as the carbon source via the traditional solid-state reaction, which presented a higher capacity of 212.06 mA h g^{-1} under 0.1 A g^{-1} upon 100 cycles compared with the bare bulk $Bi_5Nb_3O_{15}$ (figure 6.4(d)) [28]. While the same group synthesized the 1D architecture $Bi_5Nb_3O_{15}$ nanowire via the electrospinning technique without carbon-based material coating strategy [32]. And the composite exhibited a higher capacity of 267 mA h g^{-1} upon 100 cycles compared with the aforementioned $Bi_5Nb_3O_{15}$@C composite, which could be owing to tolerance of volume variation and reduced ion diffusion channels of the nanostructure. Furthermore, the reaction mechanism was explored to be a combination of conversion and alloying via the *in situ* x-ray diffraction measurements.

For $BiVO_4$, the semiconductor is useful as the anode material for lithium-ion batteries considering the acceptable band gap of 2.4 eV and layered crystal structure with multivalent bismuth and vanadium elements. Romero *et al* reported a $BiVO_4$ with fern architecture and studied its lithium-ion storage properties, which presented a capacity of 796 mA h g^{-1} under 0.12 A g^{-1}, corresponding to an ultrahigh volumetric capacity of 3984 mA h cm^{-3} [33]. Fang *et al* fabricated a $BiVO_4$ nanowire array via the hydrothermal method with uniform *in situ* polymerized polypyrrole coating on the surface of the nanowires after the cells being charged to 4 V considering the strong oxidability at this state, as shown in figure 6.4(e) [29]. The *in situ* conductive coating was beneficial for enhancing the electron transfer and protecting the composite from cracking during repeated lithiation/delithiation process, resulting in a capacity of 400 mA h g^{-1} after 50 cycles. Dubal *et al* engineered $BiVO_4$ spheres with fractal granularity containing

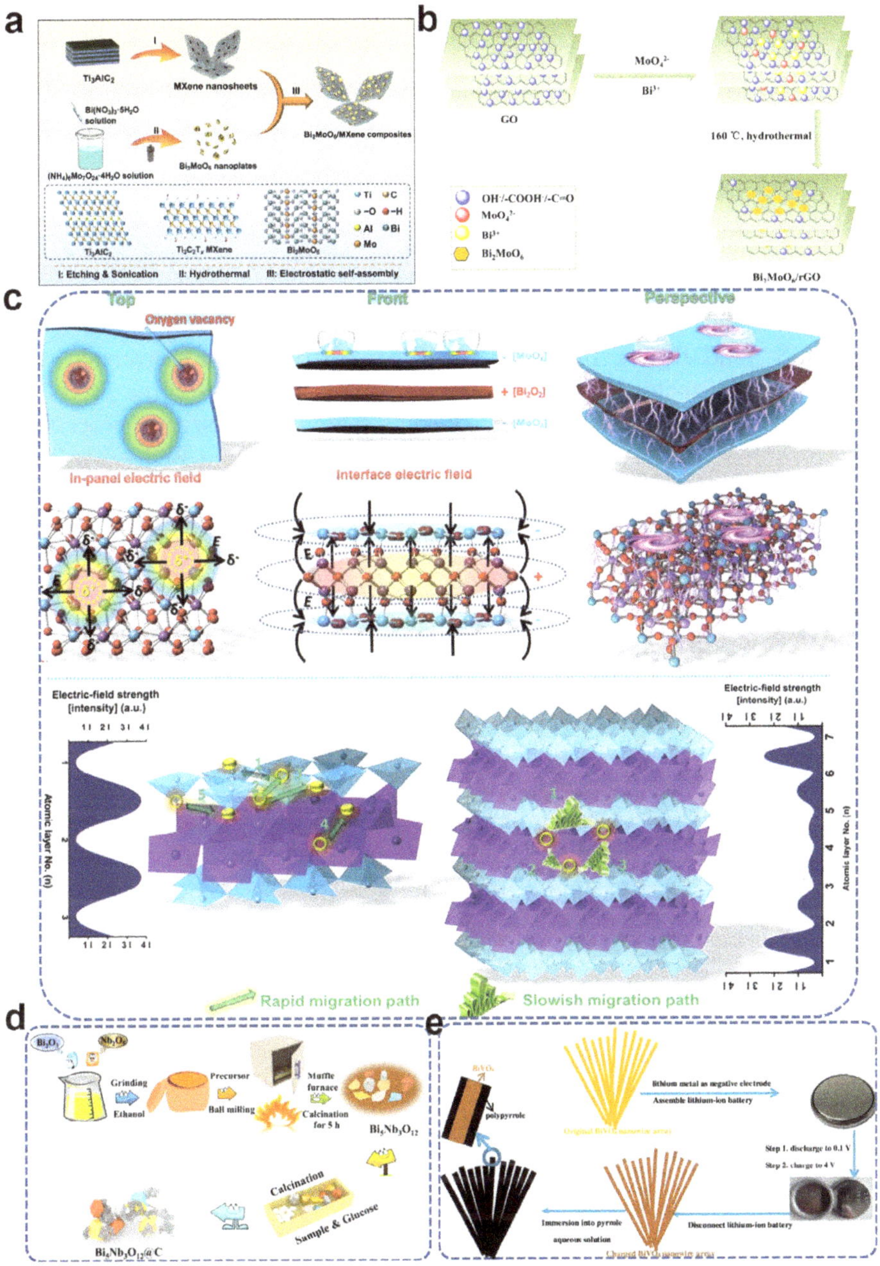

Figure 6.4. (a) Schematic illustration of the fabrication and cycling performance of the $Bi_2MoO_6/Ti_3C_2T_x$. Source: [25]. Reproduced with permission of Springer Singapore. (b) Schematic illustration of the fabrication of the Bi_2MoO_6/rGO. Source: [26]. Reproduced with permission of Elsevier Ltd. (c) Schematic illustration of the induced in-panel electric field and the interface electric field within the Bi_2MoO_6 sheets viewed along different directions, the electric field distribution and the corresponding crystal structure of the Bi_2MoO_6 sheets, and the relationship between the electric field, migration paths, and the Li transfer kinetic in Bi_2MoO_6 sheets and Bi_2MoO_6 bulk. Source: [27]. Reproduced with permission of Wiley-Blackwell. (d) Schematic illustration of the fabrication of bulk $Bi_5Nb_3O_{15}$ and $Bi_5Nb_3O_{15}$@C. Source: [28]. Reproduced with permission of Elsevier Ltd. (e) Schematic illustration of the fabrication of the $BiVO_4$ nanowires. Source: [29]. Reproduced with permission of Elsevier.

the nanoparticle with 20–40 nm in size distribution on the surface of the spheres using cost-effective surfactant-assisted hydrothermal method [34]. The $BiVO_4$ delivered a capacity of 590 mA h g^{-1} under 0.1 A g^{-1} and presented a high capacity retention of 96% after 200 cycles.

6.2.1.6 Bi anode for Na-ion batteries

For the general reaction mechanism in sodium-ion batteries, the metallic bismuth alloys three sodium ions to form the Na_3Bi involving two-phase process of Bi $\leftrightarrow$ NaBi and NaBi $\leftrightarrow$ Na_3Bi. In terms of the Bi_2X_3 compounds, including Bi_2O_3, Bi_2S_3, and Bi_2Se_3, the reaction mechanism composes conversion (for the generation of the metallic bismuth) and alloying processes (for the producing of Na_3Bi). Notably, some research indicated that there is the insertion process before the conversion reaction, corresponding to the formation of $Na_xBi_2X_3$, based on the measurement results of *in/ex situ* XRD, *in/ex situ* TEM, and so on.

6.2.1.7 Metallic bismuth-based composites

Due to the large layer spacing along the *C*-axis ($d(003) = 3.95$ Å), bismuth is also conducive to the insertion and extraction of sodium ions. And many relevant studies have reported in the field of sodium storage. In 2015, Su *et al* investigated the Bi@graphene nanocomposite as a potential anode material for the sodium-ion batteries, unveiling the sodium ion intercalation mechanism by the facile site for sodium-ion diffusion convinced by the *ex situ* XRD tests and DFT calculation [35]. Thanks to the large layer spacing and the 3D architecture of the nanocomposite, the composite exhibited high reversible capacity of 561 mA h g^{-1} within the voltage range of 0.01–2.0 V and 358 mA h g^{-1} within the range of 0.3–0.9 V. Li *et al* made a fundamental work of researching the sodium-ion storage properties of the bulk bismuth without any treatment in the glyme-based electrolytes ($NaPF_6$-diglyme electrolyte) [36]. The phase of bulk bismuth would gradually degrade into a porous integrity upon the initial cycling process, which is different from that in the carbonate-based electrolytes, as shown the schematic illustration of morphological evolution of the bismuth electrode during the initial cycles. The resulting porous structure could accelerate the diffusion of ions and increase the stability, leading to a high capacity retention of 94.4% over 2000 cycles (0.0028% capacity decay per cycle). The better performance of bismuth in glyme-based electrolytes than that in carbonate-based electrolytes can be attributed to the high wettability with the bismuth material and the stable solid electrolyte interphase (SEI) film formed in the glyme-based electrolytes, which is responsible for the high interfacial stability and facile sodium-ion diffusion. Recently, Yu's group reported an attractive bismuth-based material with 3D continuous porous structure featured with the ability of volumetric variation self-healing [37]. Notably, the interconnected bismuth ligaments offered unblocked pathway for ion/electrons and were blessed with the ability to accomplish structural recovery. As a result, the 3D porous bismuth network achieved an excellent rate performance with a capacity retention of 95.6% at a high current density of 60 A g^{-1} and a capacity of 378 mA h g^{-1} under 10 A g^{-1} after 3000 cycles, as shown in figure 6.4(b).

In order to address the severe issue of structural change, Yin *et al* synthesized the bismuth/carbon nanotubes by one-step electro-deoxidation method, as shown in figure 6.5(c) [38]. The obtained electrolytic micro-Bi/CNTs composite guaranteed the facile electron and ion transport since the carbon nanotubes played the role as the backbones and the unblocked diffusion channel networks, delivering a capacity of 381.6 mA h g^{-1} under 2 A g^{-1} with a retention of 99.5% after 1000 cycles. Zhang and Wang *et al* applied graphene nanosheets to wrap the bismuth to relieve the large volume change during the electrochemical reactions and accelerate the electron transfer. It is worth mentioning that, at an ultrahigh current density of 111 A g^{-1}, the electrode still maintained a capacity of 108.6 mA h g^{-1}, owing to the interconnected graphene served as the framework for fast sodium-ion kinetics. The operando Raman spectrum unveiled that a reversible sodium storage process also occurred on graphene nanosheet, as shown in figure 6.5(d) [39]. For the instance of heterogeneous structured pomegranate-like Bi@C nanospheres proposed by Wu's group, the pomegranate-like composite contained fine bismuth nanoparticles encapsulated into the carbon spheres, which were interlinked to form the conductive matrix to provide

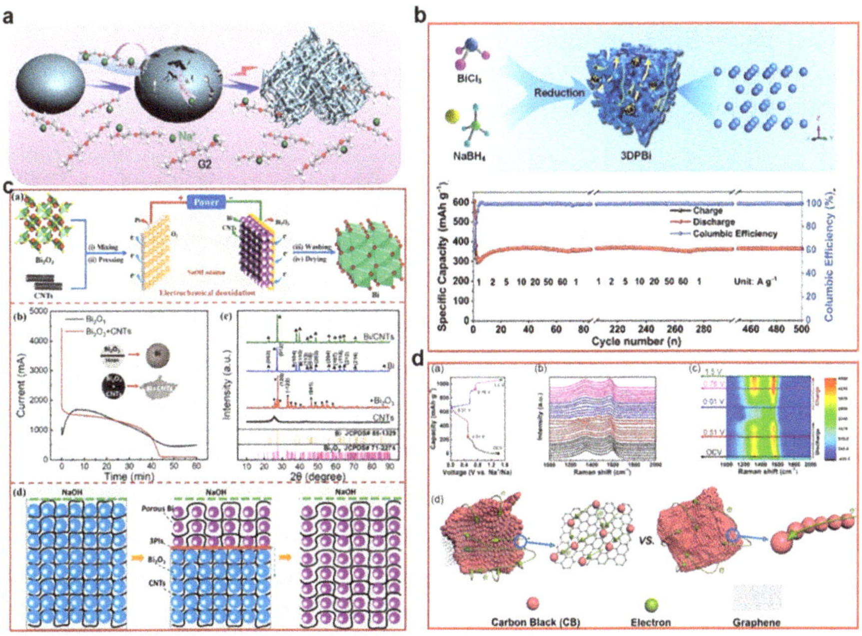

Figure 6.5. (a) Schematic illustration of morphological evolution of the bismuth electrode during cycles. Source: [36]. Reproduced with permission of Wiley-Blackwell. (b) Schematic illustration of synthesis procedure and rate performance of the 3D porous bismuth network. Source: [37]. Reproduced with permission of Wiley-VCH Verlag. (c) Schematic illustration of the electrochemical deoxidation strategy, current–time curves, XRD patterns, as well as the schematic illustration of the electro-reduction of Bi_2O_3/CNTs to Bi/CNTs. Source: [38]. Reproduced with permission of Elsevier. (d) Operando Raman characterization and schematic illustration of the effects of graphene nanosheets on the electron transport and ion storage behavior of bismuth. Source: [39]. Reproduced with permission of American Chemical Society.

the fast electron/ion pathway and prohibit the aggregation of the nanoparticles within the synthesis and electrochemical processes, enabling the composite to exhibit good cycling (340 mA h g^{-1} under 20 A g^{-1} after 1600 cycles) and rate (372.8 mA h g^{-1} under 25 A g^{-1}) performance [40].

6.2.1.8 Bismuth oxide

The sodium-ion storage performance is rare to be reported and the pioneering research work was conducted by Chandrasekaran Nithya [41]. The author chose the surfactant of cetyltrimethylammonium bromide in the synthesis which played a major role in regulating the morphology of the Bi_2O_3 nanoparticles in the range of 10–20 nm. rGO nanosheets was adopted as the substrate to support the nanoparticles and buffer the volume change upon the ion intercalation/deintercalation process. And the as-prepared composite presented a not bad rate performance. While Zhu's group [42] and Demir's group [43] applied the 1D nanotube and nanofiber structure to allow facile electron transport and alleviate the volume expansion/extraction effectively, leading to improved electrochemical performance compared to that of bare or non-modified Bi_2O_3. Considering the effect of tuning the physical/chemical properties of the heterointerface, Mai *et al* designed a heterostructured Bi_2S_3–Bi_2O_3 nanosheets, which displayed a high reversible capacity of about 630 mA h g^{-1} under 0.1 A g^{-1}, owing to the introduced built-in electric field to accelerate the ion/electron transport [44].

6.2.1.9 Bismuth sulfides

The improvement of Bi_2S_3 anode has been actively employed for sodium-ion storage taking structural modification as a prerequisite. Wrapping by graphene nanosheet is a widely applied approach to address the issues of volume expansion and slow electron transfer of Bi_2S_3. For example, Su *et al* encapsulated the Bi_2S_3 nanorods into nitrogen-doped graphene network to form the 3D targeting flexible composite with high structural stability and unblocked ion/electron diffusion channels, presenting a high capacity of 649 mA h g^{-1} under 62.5 mA g^{-1} (figure 6.6(a)). Moreover, the defects generated from the nitrogen doping offered extra anchoring active sites for sodium-ion storage. Furthermore, the combined conversion ($Bi_2S_3 \leftrightarrow$ Bi) and alloying (Bi $\leftrightarrow$ NaBi $\leftrightarrow$ Na_2Bi $\leftrightarrow$ Na_3Bi) mechanism was illustrated by the *ex situ* XRD measurement. Li *et al* anchored the Bi_2S_3 nanorods on the rGO nanosheet, which were interconnected with the C–S bonds. The synergetic effects of the chemical bonds and different component enabled the composite to deliver a capacity of 226 mA h g^{-1} under 0.1 A g^{-1} after 100 cycles [45].

While Choi *et al* encapsulated the Bi_2S_3 into the carbon spheres with the yolk–shell structure via the sulfur-impregnation procedure using glucose as the carbon source. The carbon shells ensured the durable structure and long-term cycling performance with capacity of 282.4 mA h g^{-1} under 0.2 C after 300 cycles and 413.0 mA h g^{-1} under 10 C, as shown in figure 6.6(b) [46]. The previous report indicated that the minimal doping of selenium into Bi_2S_3 could enhance the electrical conductivity and generate the selenium-containing inactive phase to improve the electrochemical property [47, 48]. Zhang *et al* designed Bi_2S_3 nanoneedles grown on

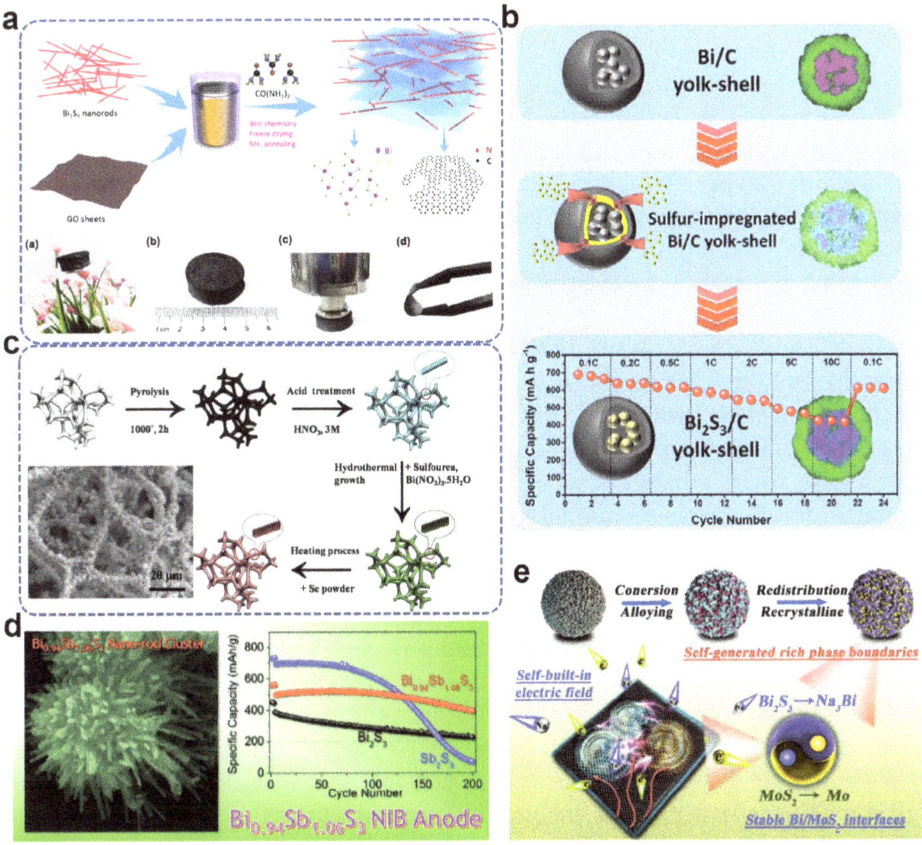

Figure 6.6. (a) Schematic illustration of the synthesis of the Bi_2S_3/N-doped graphene and the lightweight and flexible nature of the composite. Source: [44]. Reproduced with permission of American Chemical Society. (b) Schematic illustration of morphological evolution of the yolk–shell structured Bi_2S_3@C. Source: [46]. Reproduced with permission of Elsevier. (c) Synthesis schematic illustration of Se–Bi_2S_3/CF. Source: [49]. Reproduced with permission of Elsevier B.V. (d) Schematic illustration of morphological evolution and the rate capability of the $Bi_{0.94}Sb_{1.06}S_3$–graphite composite. Source: [50]. Reproduced with permission of American Chemical Society. (e) Schematic illustration of the synthesis of the Bi_2S_3/MoS_2 heterostructure and its merits. Source: [51]. Reproduced with permission of Wiley-VCH Verlag.

the surface of the carbon foam, which was subjected to a selenizing process to obtained the Se–Bi_2S_3/CF, and the synthesis schematic illustration is shown in figure 6.6(c) [49]. The carbon network played the role in suppressing the volume change, assuring the fast ion diffusion and preventing the energy density from degradation. As a result, the as-prepared Se–Bi_2S_3/CF displayed a 441.8 mA h g^{-1} capacity after 100 cycles with the Coulombic efficiency of approximating 100%.

Heterostructure designing was also applied to boost the sodium-ion storage for Bi_2S_3. For example, Manthiram's group fabricated a $Bi_{0.94}Sb_{1.06}S_3$–graphite composite composed of binary counterparts Bi_2S_3 and Sb_2S_3 via the hydrothermal reaction. The synergistic effect between the counterparts made the composite exhibit good cycling performance with a capacity retention of 79% over 200 cycles within the voltage range of 0.01–2.8 V (2.8–0.85 V for the conversion reaction and 0.85–0.01 V for the alloying reaction), as shown in figure 6.6(d) [50]. While Yang *et al* synthesized a Bi_2S_3/MoS_2 heterogeneous material with abundant phase boundaries, which introduced the interior self-built-in electric-field to boost the electron/ion transfer. Notable, the stable Bi/Na_2S interface could improve the conversion/alloying reversibility and keep integrity without agglomeration and pulverization, leading to a good long-term cycling performance of a capacity 323.4 mA h g^{-1} after 1200 cycles under 10 A g^{-1}, as shown in figure 6.6(e) [51].

6.2.1.10 Bismuth selenides and tellurides

There are only a few reports on the sodium-ion storage performance of Bi_2Se_3, which might can be attributed to the difficult synthesis process. Zhou and Chen *et al* reported a Bi_2Se_3/graphene composite fabricated via a selenization treatment of the Bi/graphene intermediate. A series of *ex situ* measurements further proved the combination of conversion and alloying mechanism. The uniformly distribution of nanoparticles in size of about 100 nm on the graphene sheets connected by the interfacial interaction could effectively restrain the volume change and provide increased electrical conductivity, showing a remarkable rate capability of 183 mA h g^{-1} at 10 A g^{-1} [52]. Chen *et al* synthesized a Bi_2Se_3/C composite using graphite as carbon source by a simple ball-milling method. And composite delivered a capacity of 469 mA h g^{-1} at 0.1 A g^{-1} after 100 cycles [53]. While Chen and Han *et al* reported a topological insulator of $MoSe_2/Bi_2Se_3$ heterostructures (growing $MoSe_2$ nanosheets onto Bi_2Se_3 nanoflakes) with fast interface electrical conductivity and large interlayer spacing, which enabled the composite to achieve a high-rate capacity of 244 mA h g^{-1} at 10 A g^{-1} [54].

For the bismuth tellurides (Bi_2Te_3), its unique nature of topological insulator has attracted increased attention due to its free electron transfer along the 2D surface of the material regardless of the relatively low theoretical capacity compared with its sulfide and selenide counterparts. Jia and Zhou *et al* synthesized a layered Bi_2Te_3 nanoplate with large interlay spacing of 1.02 nm attached on graphene (figure 6.7(a)) [55]. The as-obtained composite achieved a high-rate capability of 203 mA h g^{-1} at 5 A g^{-1} owing to the synergistic effect between Bi_2Te_3 and graphene, quintuple layer sandwich structure of Bi_2Te_3, and its higher density. Zhao *et al* encapsulated the Bi_2Te_3 into the MOFs derived carbon framework via a solvothermal process, as shown in figure 6.7(b) [56]. The carbon network could effectively tolerate the volume variations and accelerate the electron transfer, enabling the composite to exhibit a capacity of 236.4 mA h g^{-1} at 100 mA g^{-1} after 100 cycles. Sharma's group reported tetradymite-type Bi_2Te_3 synthesized via a solid-state method and studied the lithium, sodium, and potassium storage properties, revealing the inferior cycling performance in sodium and potassium storage with more obvious capacity degradation due to the larger volume change [58]. Li's group covered the PPy layer on the surface of Bi_2Te_3 to form a composite of Bi_2Te_3@PPy, in which the PPy coating could protect the structural integrity and prevent the structural pulverization. The composited exhibited enhanced rate capability and cycling performance [57]. And the specific

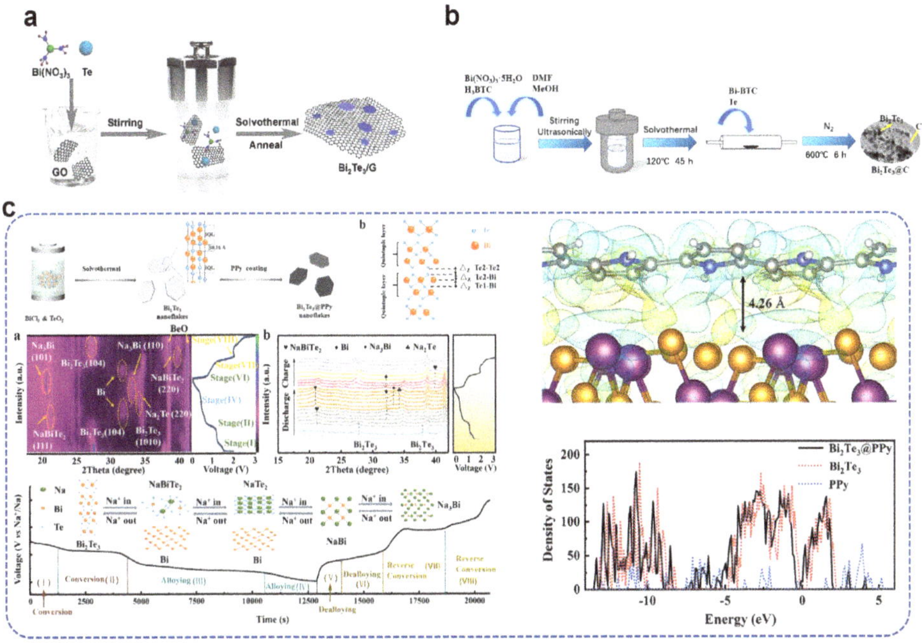

Figure 6.7. Schematic illustrations of the synthesis of (a) the Bi_2Te_3 nanoplate on graphene. Source: [55]. Reproduced with permission of Royal Society of Chemistry, and (b) Bi_2Te_3@C. Source: [56]. Reproduced with permission of Elsevier B.V. (c) Synthesis schematic illustration, *in situ* XRD tests, schematic illustration of morphological evolution, and DFT calculation results of Bi_2Te_3@PPy. Source: [57]. Reproduced with permission of Elsevier.

phase evolution was sufficiently investigated by the *in/ex situ* XRD measurements and *ex situ* TEM observation. Furthermore, the DFT calculations revealed that the heterointerface between Bi_2Te_3 and PPy interlayer was beneficial for the sodium-ion fast transfer, and the narrowing band gap would enhance the conductivity of the system, as shown in figure 6.7(c).

6.2.2 Bi anode for K-ion batteries

Due to the low standard electrode potential (K/K^+: −2.93 V) and low cost of potassium source, PIBs have appeared as the competitive energy storage system. Therefore, bismuth-based materials have gathered extensive interests as the potassium anode materials. And their large volume change makes the rational design to be taken into account in the structural fabrication, morphology engineering, electrolyte modification, as so on. Zhang *et al* gave a clear understanding of the phase transitions of K–Bi by the electrochemical profiles, *in situ* XRD measurements, and DFT calculation, as shown in figure 6.8. The initial discharge profile showed only one plateau, which was different from that in the following cycles. And the sequence alloying process obeyed the Bi ↔ KBi_2 ↔ K_3Bi_2 ↔ K_3Bi procedures [59]. While Guo *et al* considered the intercalation of potassium into bismuth underwent a solid-solution reaction of Bi ↔ Bi(K) and Bi(K) ↔ K_5Bi_4 ↔ K_3Bi by the *ex situ* XRD tests

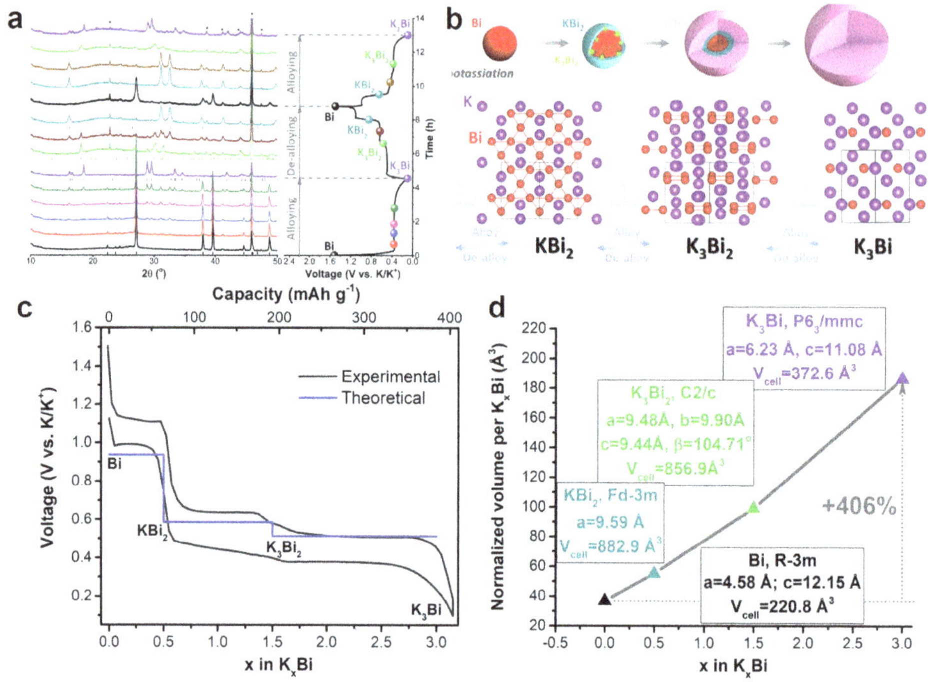

Figure 6.8. (a) *In situ* XRD patterns, (b) alloying and dealloying processes in bismuth electrode in the first discharge and the following cycles, (c) the calculated electrochemical profile compared with the second discharge/charge curves, (d) volume change of the identified K–Bi alloys. Source: [59]. Reproduced with permission of Wiley-VCH Verlag.

[60]. And the authors believed that the bis(fluorosulfonyl)imide (KFSI) salt favored to form a stable, thin, robust SEI layer on the surface of the electrode.

6.2.2.1 Metallic bismuth-based composites

In order to alleviate the mechanical strain produced during the potassium ion intercalation/extraction, Yu's group designed two metallic bismuth/carbon composites of multicore–shell Bi@N-doped carbon nanospheres (Bi@N–C) (figure 6.9(a)) [61] and bismuth in the 3D interconnected macroporous graphene framework (Bi@3DGFs) (figure 6.9(b)) [62]. The carbon coating and graphene conductive carbon network, without exception, took the part in preventing volume swelling/shrinkage, decreasing the diffusion transportation of ions and electrons, increasing the electrical conductivity of the whole composite, and avoiding the SEI layer formation directly on the bismuth particles. Therefore, both samples exhibited excellent rate capability and cycling performance. It should be pointed out that the Bi@N–C displayed an ultrahigh rate performance with a reversible capacity of 152 mA h g^{-1} at 100 A g^{-1} and long-term cycle life with a capacity of 203 mA h g^{-1} at 10 A g^{-1} after 1000 cycles. Guan *et al* fabricated a multicore–shell structured Bi@N–C composite containing bismuth nanoparticles sealed within the carbon sheets with large interior free space via the self-template approach, as shown in

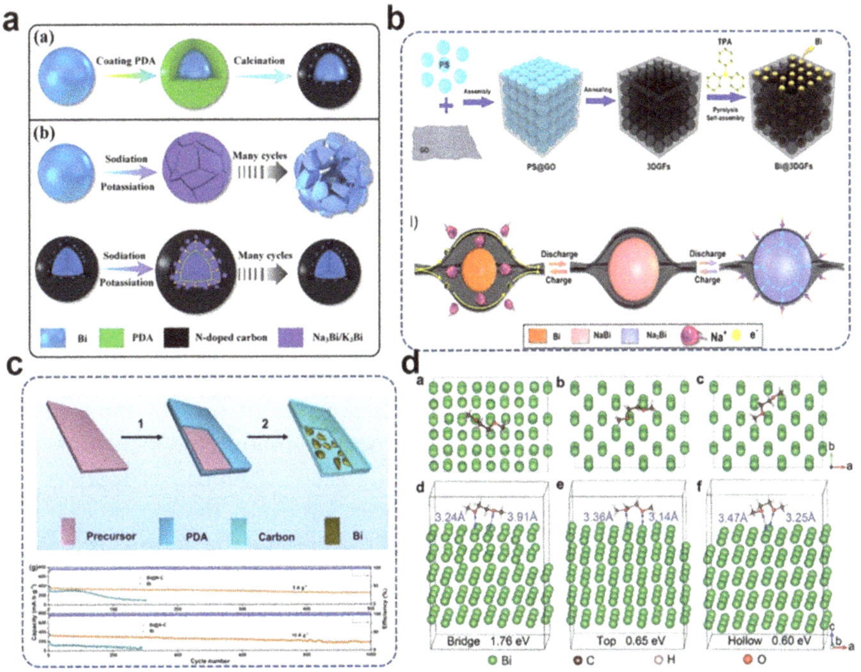

Figure 6.9. (a) Schematic illustration of the synthesis and degradation mechanisms and morphology evolution of the charge/discharge process (compared with bare bismuth) of the Bi@N–C. Source: [61]. Reproduced with permission of Wiley-VCH Verlag. (b) Schematic illustration of the synthesis of Bi@3DGF. Source: [62]. Reproduced with permission of Royal Society of Chemistry. (c) Schematic illustration of the synthesis and cycling performance of Bi@N–C. Source: [63]. Reproduced with permission of Royal Society of Chemistry. (d) Schematic illustration of the different adsorption models of DME molecules on the (012) plane. Source: [64]. Reproduced with permission of John Wiley and Sons Ltd.

figure 6.9(c). The large fraction pseudocapacitive contribution and rapid diffusion of potassium ion and electrons enabled the composite to exhibit long-term cycling durability with a capacity up to 302 mA h g^{-1} at 1 A g^{-1} over 1000 cycles [63]. Li and Wang *et al* pointed out by the electrochemical properties evaluation and DFT calculation that the bismuth would gradually transfer to 3D porous network, realizing rapid ion diffusion and structural confinement, and the formed pores linked the bismuth atoms with the solvent dimethoxyethane (DME) molecules. Figure 6.9(d) illustrates the different adsorption models of DME molecules on the (012) plane, and the lowest adoption indicates that the chemical adsorption of the DME molecules on the bismuth surface induced the movement of its surface atoms, and further favored the formation of the 3D porous structure [64].

6.2.2.2 Bi_2X_3 (X = O, S, Se, Te)

So far, the application of Bi_2X_3 in PIBs has been rarely reported. Liu and Wang *et al* reported a composite of $BiSb@Bi_2O_3/SbO_x@C$ using the template-assisted *in situ* pyrogenic decomposition approach (figure 6.10(a)) by application of BiSb alloy to increase the electrical conductivity for the composite. Combining the porous carbon network with the synergistic effect between bismuth and antimony, the composited delivered a capacity of 214 mA h g^{-1} at 1 A g^{-1} over 500 cycles [65]. Li and Du *et al* fabricated the Bi_2S_3@rGO composite by the visible-light-assisted approach (figure 6.10(b)) and sheds light on the multi-step conversion-alloying reaction mechanism. The composite exhibited a capacity of 237 mA h g^{-1} under 2 A g^{-1} over 300 cycles [66]. Moreover, the *ex situ* measurement was applied to validate the conversion and alloying potassium mechanism. Wu *et al* synthesized the N-doped rGO-Bi_2S_3 taking into consideration of restraining the 'shuttle effect' and thus capacity degradation. The N-doping was believed to introduce the C–S bonds to confine the polysulfide formed within electrochemical reaction and prevent the capacity from fading (figure 6.10(c)). Combining the nitrogen doping with other advantages, such as more adsorption sites for potassium ions and increased electrical conductivity, the composite exhibited good electrochemical properties with a capacity of 331.9 mA h g^{-1} under 5 A g^{-1} over 500 cycles [67].

Heterostructure designing was also applied to boosting the potassium ion storage for Bi_2S_3. Li's group designed the two heterostructured Bi_2S_3/MoS_2 encapsulated in nitrogen-doped carbon shell (BMS@NC) and Sb_2S_3–Bi_2S_3@C@rGO composites (figure 6.10(d)) considering the advantage of the interior self-built-in electric-field to boost the electron/ion transfer. Both composites exhibited improved potassium storage properties compared with the corresponding single phase metal sulfides [68, 69]. For the Bi_2Se_3, Zheng *et al* synthesized the rod-like Bi_2Se_3@C architecture using MOFs as the precursor via the *in situ* selenization method. The porous carbon framework could absorb the volumetric swelling, prevent the aggregation of Bi_2Se_3 nanoparticles, increase electrical conductivity, and shorten the diffusion pathway for potassium ions, leading to a good electrochemical performance with a capacity of 543 mA h g^{-1} under 1 A g^{-1} over 5000 cycles, as shown in figure 6.10(e). Moreover, the multi-step conversion-alloying reaction mechanism was verified by the *ex situ* XRD tests [70]. In addition, Tuan *et al* wrapped the Bi_2Se_3 nanosheets on the reduced graphene oxide and the exterior N-doping carbon coating (Bi_2Se_3@NC@rGO), which was proved to display good electrochemical properties due to the formed strong C–O–Bi bonding to reinforce the structural integrity and reaction kinetics. Notable, the reaction mechanism was investigated by the *in situ* XRD measurements and *ex situ* TEM observation (figure 6.10(f)), unveiling that a conversion-alloying reaction involved 12-electron transfer per formula unit, and the K_2Se product of conversion reaction can efficiently suppress the volume expansion/extraction. The obtained composite exhibited a cycling life of 612 mA h g^{-1} under 0.1 A g^{-1} over 1000 cycles [71].

6.3 Challenge of Bi anode for Li, Na, K-ion batteries

Actually, the electrolyte plays an important role in regulation of the formation of SEI layer, which is the key factor to access durable cycling performance. As mentioned above, the glyme-based electrolytes are particularly suitable for the

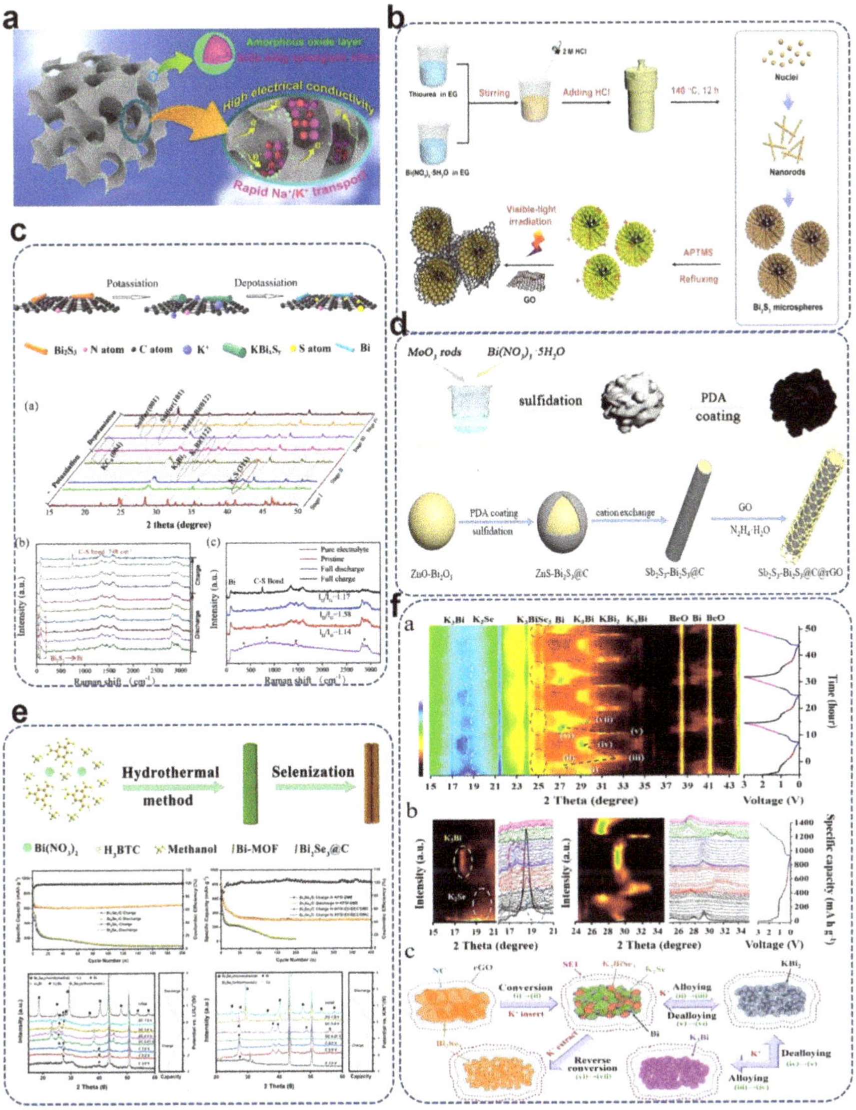

Figure 6.10. (a) The structural merit of $BiSb@Bi_2O_3/SbO_x@C$ for fast ion transport. Source: [65]. Reproduced with permission of Academic Press Inc. (b) Schematic illustration of synthesis and morphology evolution of the $Bi_2S_3@rGO$ composite. Source: [66]. Reproduced with permission of Royal Society of Chemistry. (c) Schematic illustration of synthesis and reaction mechanism of N-doped rGO-Bi_2S_3. Source: [67]. Reproduced with permission of Elsevier B.V. (d) Schematic illustration of synthesis of the BMS@NC and Sb_2S_3-$Bi_2S_3@C@rGO$. Source: [68, 69]. Reproduced with permission of American Chemical Society and John Wiley and Sons Ltd. (e) Schematic illustration of synthesis, cycling performance, *ex situ* XRD patterns of the $Bi_2Se_3@C$ composite. Source: [70]. Reproduced with permission of American Chemical Society. (f) *In situ* XRD patterns and schematic illustration of phase evolution of the $Bi_2Se_3@NC@rGO$ composite. Source: [71]. Reproduced with permission of Elsevier.

interfacial stability and facile sodium-ion diffusion due to elasticity and stronger binding affinity by alkoxy (O–K) edge groups compared with carbonate-derived electrolytes. The resulting SEI layer was flexible to tolerant to the huge volume expansion and avoid the irreversible loss of the internal active materials to form fresh electrode–electrolyte interfaces. Expect of the solvent in the electrolyte, it is reported that the concentrated electrolyte containing the potassium salt, potassium bis(fluorosulfonyl)imide (KFSI), is effective to generate a thin and robust SEI layer to restrain the capacity degradation of the composite [72]. However, the suitable potassium salts are usually expensive, not to mention the large dose, causing the rising costs. Therefore, it is significant to select suitable and low-cost electrolytes taking into account to achieve satisfied electrochemical performance and reasonable price. Moreover, comprehensive study of the interface between bismuth-based material and electrolytes is needed, including composition and structure, ion transfer, effective SEI film formation, interface stability and so on. Furthermore, it is still a challenge for the bismuth-based electrode material in the structural engineering, morphology design, employing theoretical calculation, and advanced characterization techniques to unveil the reaction mechanism, and choosing matched cathode material for the full-cells, and so on, which require sufficient research.

References

[1] Lan J-L, Jin Y, Qin C, Yu Y and Yang X 2017 Bio-inspired rose-like Bi@nitrogen-enriched carbon towards high-performance lithium-ion batteries *ChemistrySelect* **2** 7178–84

[2] Hong W, Wang A, Li L, Qiu T, Li J, Jiang Y, Zou G, Peng H, Hou H and Ji X 2021 Bi dots confined by functional carbon as high-performance anode for lithium ion batteries *Adv. Funct. Mater.* **31** 2000756

[3] Qiu X, Wang B, Jing P, Zhang Y, Zhang M, Wang Q, Chen X, Zhang Y and Wu H 2020 Bismuth dots imbedded in ultralong nitrogen-doped carbon tubes for highly efficient lithium ion storage *Inorg. Chem. Front.* **7** 4854–64

[4] Zhong Y, Li B, Li S, Xu S, Pan Z, Huang Q, Xing L, Wang C and Li W 2018 Bi Nanoparticles anchored in N-doped porous carbon as anode of high energy density lithium ion battery *Nano-Micro Letters* **10** 56

[5] Hong W, Ge P, Jiang Y, Yang L, Tian Y, Zou G, Cao X, Hou H and Ji X 2019 Yolk–shell-structured bismuth@N-doped carbon anode for lithium-ion battery with high volumetric capacity *ACS Appl. Mater. Interfaces* **11** 10829–40

[6] Yuan H, Jin Y, Chen X, Lan J, Yu Y and Yang X 2019 Large-scale fabrication of egg-carton-inspired Bi/C composite toward high volumetric capacity and long-life lithium ion batteries *ACS Sustain. Chem. Eng.* **7** 6033–42

[7] Jin Y, Yuan H, Lan J-L, Yu Y, Lin Y-H and Yang X 2017 Bio-inspired spider-web-like membranes with a hierarchical structure for high performance lithium/sodium ion battery electrodes: the case of 3D freestanding and binder-free bismuth/CNF anodes *Nanoscale* **9** 13298–304

[8] Xiong P *et al* 2018 Two-dimensional unilamellar cation-deficient metal oxide nanosheet superlattices for high-rate sodium ion energy storage *ACS Nano* **12** 12337–46

[9] Li Y, Trujillo M A, Fu E, Patterson B, Fei L, Xu Y, Deng S, Smirnov S and Luo H 2013 Bismuth oxide: a new lithium-ion battery anode *J. Mater. Chem.* A **1** 12123–27

[10] Mei J, Liao T, Ayoko G A and Sun Z 2019 Two-dimensional bismuth oxide heterostructured nanosheets for lithium- and sodium-ion storages *ACS Appl. Mater. Interfaces* **11** 28205–12

[11] Fang W, Zhang N, Fan L and Sun K 2016 Bi2O3 nanoparticles encapsulated by three-dimensional porous nitrogen-doped graphene for high-rate lithium ion batteries *J. Power Sources* **333** 30–6

[12] Prasath A, Athika M, Duraisamy E, Selva Sharma A, Sankar Devi V and Elumalai P 2019 Carbon quantum dot-anchored bismuth oxide composites as potential electrode for lithium-ion battery and supercapacitor applications *ACS Omega* **4** 4943–54

[13] Liu H, Yang M, Yi Z, Duan T and Yao W 2021 Bi2O3/Bi nanocomposites confined by N-doped honeycomb-like porous carbon for high-rate and long-life lithium storage *Appl. Mater. Today* **22** 100885

[14] Ni J, Bi X, Jiang Y, Li L and Lu J 2017 Bismuth chalcogenide compounds Bi2×3 (X=O, S, Se): applications in electrochemical energy storage *Nano Energy* **34** 356–66

[15] Yu Y, Jin C H, Wang R H, Chen Q and Peng L M 2005 High-quality ultralong Bi2S3 nanowires: structure, growth, and properties *J. Phys. Chem.* B **109** 18772–76

[16] Zhao Y, Gao D, Ni J, Gao L, Yang J and Li Y 2014 One-pot facile fabrication of carbon-coated Bi2S3 nanomeshes with efficient Li-storage capability *Nano Res.* **7** 765–73

[17] Liang H, Ni J and Li L 2017 Bio-inspired engineering of Bi2S3-PPy yolk–shell composite for highly durable lithium and sodium storage *Nano Energy* **33** 213–20

[18] Li Z, He M, Bo B, Wei H, Liu Y, Wen H, Liu Y, Zhang K, Zhang P and Li B 2021 Bi2S3 nanorods hosted on rGO sheets from pyrolysis of molecular precursors for efficient Li-ion storage *Energy Environ. Mater.* **4** 577–85

[19] Xiang J, Liu Z and Song T 2018 Sandwich-like graphene–Bi2S3 hybrid derived from (BiO) 2CO3 nanosheets as advanced anode materials for lithium/sodium ion batteries *J. Alloys Compd.* **768** 426–32

[20] Wang S, Li W, Song H, Mao C, Zhang Z, Peng H and Li G 2019 Nitrogen-enriched carbon-coated flower-like bismuth sulfide architectures towards high-performance lithium-ion battery anodes *Inorg. Chem. Front.* **6** 1275–81

[21] Chai W, Yang F, Yin W, You S, Wang K, Ye W, Rui Y and Tang B 2019 Bi2S3/C nanorods as efficient anode materials for lithium-ion batteries *Dalton Trans.* **48** 1906–14

[22] Wang L, Han Y, Feng X, Zhou J, Qi P and Wang B 2016 Metal–organic frameworks for energy storage: batteries and supercapacitors *Coord. Chem. Rev.* **307** 361–81

[23] Chen X, Tang H, Huang Z, Zhou J, Ren X, Huang K, Qi X and Zhong J 2017 Flexible bismuth selenide/graphene composite paper for lithium-ion batteries *Ceram. Int.* **43** 1437–42

[24] Jin R, Sun M and Li G 2017 CNTs@C@Bi2Se3 composite as an improved-performance anode for lithium ion batteries *Ceram. Int.* **43** 17093–99

[25] Zhang P, Wang D, Zhu Q, Sun N, Fu F and Xu B 2019 Plate-to-layer Bi2MoO6/MXene-heterostructured anode for lithium-ion batteries *Nano-Micro Lett.* **11** 81

[26] Zhai X, Gao J, Xue R, Xu X, Wang L, Tian Q and Liu Y 2018 Facile synthesis of Bi2MoO6/reduced graphene oxide composites as anode materials towards enhanced lithium storage performance *J. Colloid Interface Sci.* **518** 242–51

[27] Zheng Y, Zhou T, Zhao X, Pang W K, Gao H, Li S, Zhou Z, Liu H and Guo Z 2017 Atomic interface engineering and electric-field effect in ultrathin Bi2MoO6 nanosheets for superior lithium ion storage *Adv. Mater.* **29** 1700396

[28] Li Y, Zheng R, Yu H, Cheng X, Zhu H, Bai Y, Liu T, Shui M and Shu J 2018 Carbon-coated Bi5Nb3O15 as anode material in rechargeable batteries for enhanced lithium storage *Ceram. Int.* **44** 11505–11

[29] Fang D, Cui M, Bao R, Yi J and Luo Z 2020 In-situ coating polypyrrole on charged BiVO4 nanowire arrays to improve lithium-ion storage properties *Solid State Ionics* **346** 115222

[30] Zhu Y, Cao C, Zhang J and Xu X 2015 Two-dimensional ultrathin ZnCo2O4 nanosheets: general formation and lithium storage application *J. Mater. Chem.* A **3** 9556–64

[31] Tang Q, Zhou Z and Shen P 2012 Are MXenes promising anode materials for li ion batteries? Computational studies on electronic properties and Li storage capability of Ti3C2 and Ti3C2X2 (X = F, OH) monolayer *JACS* **134** 16909–16

[32] Li Y, Zheng R, Yu H, Cheng X, Liu T, Peng N, Zhang J, Shui M and Shu J 2019 Fabrication of one-dimensional architecture Bi5Nb3O15 nanowires by electrospinning for lithium-ion batteries with enhanced electrochemical performance *Electrochim. Acta* **299** 894–901

[33] Dubal D P, Patil D R, Patil S S, Munirathnam N R and Gomez-Romero P 2017 BiVO4 fern architectures: a competitive anode for lithium-ion batteries *ChemSusChem.* **10** 4163–69

[34] Patil D R, Jadhav S D, Mungale A, Kalekar A S and Dubal D P 2019 Fractal granular BiVO4 microspheres as high performance anode material for Li-ion battery *Mater. Lett.* **252** 235–38

[35] Su D, Dou S and Wang G 2015 Bismuth: a new anode for the Na-ion battery *Nano Energy* **12** 88–95

[36] Wang C, Wang L, Li F, Cheng F and Chen J 2017 Bulk bismuth as a high-capacity and ultralong cycle-life anode for sodium-ion batteries by coupling with glyme-based electrolytes *Adv. Mater.* **29** 1702212

[37] Cheng X, Shao R, Li D, Yang H, Wu Y, Wang B, Sun C, Jiang Y, Zhang Q and Yu Y 2021 A self-healing volume variation three-dimensional continuous bulk porous bismuth for ultrafast sodium storage *Adv. Funct. Mater.* **31** 2011264

[38] Hu Z, Li X, Qu J, Zhao Z, Xie H and Yin H 2021 Electrolytic bismuth/carbon nanotubes composites for high-performance sodium-ion battery anodes *J. Power Sources* **496** 229830

[39] Xu C, Gong Y, Cao G, Wei H, Zhang X, Yin F and Wang G 2020 Balancing gravimetric and volumetric performances of microsized Bi/graphene anode toward practical sodium ion storage *ACS Sustain. Chem. Eng.* **8** 17327–34

[40] Zhu J, Wang J, Li G, Huang L, Cao M and Wu Y 2020 Heterogeneous structured pomegranate-like Bi@C nanospheres for high-performance sodium storage *J. Mater. Chem.* A **8** 25746–55

[41] Nithya C 2015 Bi2O3@reduced graphene oxide nanocomposite: an anode material for sodium-ion storage *ChemPlusChem.* **80** 1000–6

[42] Yin H, Cao M-L, Yu X-X, Zhao H, Shen Y, Li C and Zhu M-Q 2017 Self-standing Bi2O3 nanoparticles/carbon nanofiber hybrid films as a binder-free anode for flexible sodium-ion batteries *Mater. Chem. Front.* **1** 1615–21

[43] Demir E, Hayat Soytas S and Demir-Cakan R 2019 Bismuth oxide nanoparticles embedded carbon nanofibers as self-standing anode material for Na-ion batteries *Solid State Ionics* **342** 115066

[44] Luo W, Li F, Li Q, Wang X, Yang W, Zhou L and Mai L 2018 Heterostructured Bi2S3–Bi2O3 nanosheets with a built-in electric field for improved sodium storage *ACS Appl. Mater. Interfaces* **10** 7201–7

[45] Li S, Gao C, Hua D, Wang G, Guo S, Qiu J and Su X 2019 Bi2S3 nanorods bonding on reduced graphene oxide surface as advanced anode materials for sodium-ion batteries *Energy Technol.* **7** 1800876

[46] Kim H, Kim D, Lee Y, Byun D, Kim H-S and Choi W 2020 Synthesis of Bi2S3/C yolk–shell composite based on sulfur impregnation for efficient sodium storage *Chem. Eng. J.* **383** 123094

[47] Xin S, Yu L, You Y, Cong H-P, Yin Y-X, Du X-L, Guo Y-G, Yu S-H, Cui Y and Goodenough J B 2016 The electrochemistry with lithium versus sodium of selenium confined to slit micropores in carbon *Nano Lett.* **16** 4560–68

[48] Klavetter K C, Pedro de Souza J, Heller A and Mullins C B 2015 High tap density microparticles of selenium-doped germanium as a high efficiency, stable cycling lithium-ion battery anode material *J. Mater. Chem.* A **3** 5829–34

[49] Chen X, Hong Y, Ge X, Li C, Miao X, Wang P, Zhang Z and Yin L 2020 Se-doped Bi2S3 nanoneedles grown on the three-dimensional carbon foam as a self-supported anode for high-performance sodium ion batteries *J. Alloys Compd.* **825** 153901

[50] Zhao Y and Manthiram A 2015 Bi0.94Sb1.06S3 nanorod cluster anodes for sodium-ion batteries: enhanced reversibility by the synergistic effect of the Bi2S3–Sb2S3 solid solution *Chem. Mater.* **27** 6139–45

[51] Cao L, Liang X, Ou X, Yang X, Li Y, Yang C, Lin Z and Liu M 2020 Heterointerface engineering of hierarchical Bi2S3/MoS2 with self-generated rich phase boundaries for superior sodium storage performance *Adv. Funct. Mater.* **30** 1910732

[52] Li D, Zhou J, Chen X and Song H 2018 Graphene-loaded Bi2Se3: a conversion–alloying-type anode material for ultrafast gravimetric and volumetric Na storage *ACS Appl. Mater. Interfaces* **10** 30379–87

[53] Xie L, Yang Z, Sun J, Zhou H, Chi X, Chen H, Li A X, Yao Y and Chen S 2018 Bi2Se3/C nanocomposite as a new sodium-ion battery anode material *Nano-Micro Lett.* **10** 50

[54] Li Y, Han M, Zhou Z, Xia X, Chen Q and Chen M 2021 Topological insulator-assisted MoSe2/Bi2Se3 heterostructure: achieving fast reaction kinetics toward high rate sodium-ion batteries *ChemElectroChem.* **8** 697–704

[55] Sun D, Zhang G, Li D, Liu S, Jia X and Zhou J 2019 A layered Bi2Te3 nanoplates/graphene composite with high gravimetric and volumetric performance for Na-ion storage *Sustain. Energy Fuels* **3** 3163–71

[56] Wen S, Zhao J, Zhu Y, Mao J, Wang H and Xu J 2020 Carbon-encapsulated Bi2Te3 derived from metal–organic framework as anode for highly durable lithium and sodium storage *J. Alloys Compd.* **837** 155536

[57] Liu X, Si Y, Li K, Xu Y, Zhao Z, Li C, Fu Y and Li D 2021 Exploring sodium storage mechanism of topological insulator Bi2Te3 nanosheets encapsulated in conductive polymer *Energy Storage Mater.* **41** 255–63

[58] Gillard C H R, Jana P P, Rawal A and Sharma N 2021 Electrochemical phase evolution of tetradymite-type Bi2Te3 in lithium, sodium and potassium ion half cells *J. Alloys Compd.* **854** 155621

[59] Huang J, Lin X, Tan H and Zhang B 2018 Bismuth microparticles as advanced anodes for potassium-ion battery *Adv. Energy Mater.* **8** 1703496

[60] Zhang Q, Mao J, Pang W K, Zheng T, Sencadas V, Chen Y, Liu Y and Guo Z 2018 Boosting the potassium storage performance of alloy-based anode materials via electrolyte salt chemistry *Adv. Energy Mater.* **8** 1703288

[61] Yang H, Xu R, Yao Y, Ye S, Zhou X and Yu Y 2019 Multicore–shell Bi@N-doped carbon nanospheres for high power density and long cycle life sodium- and potassium-ion anodes *Adv. Funct. Mater.* **29** 1809195

[62] Cheng X, Li D, Wu Y, Xu R and Yu Y 2019 Bismuth nanospheres embedded in three-dimensional (3D) porous graphene frameworks as high performance anodes for sodium- and potassium-ion batteries *J. Mater. Chem.* A **7** 4913–21

[63] Shi X, Zhang J, Yao Q, Wang R, Wu H, Zhao Y and Guan L 2020 A self-template approach to synthesize multicore–shell Bi@N-doped carbon nanosheets with interior void space for high-rate and ultrastable potassium storage *J. Mater. Chem.* A **8** 8002–9

[64] Lei K, Wang C, Liu L, Luo Y, Mu C, Li F and Chen J 2018 A Porous network of bismuth used as the anode material for high-energy-density potassium-ion batteries *Angew. Chem. Int. Ed.* **57** 4687–91

[65] Wang Z, Duan C, Wang D, Dong K, Luo S, Liu Y, Wang Q, Zhang Y and Hao A 2020 BiSb@Bi2O3/SbOx encapsulated in porous carbon as anode materials for sodium/potassium-ion batteries with a high pseudocapacitive contribution *J. Colloid Interface Sci.* **580** 429–38

[66] Liu Y, Li M, Zheng Y, Lin H, Wang Z, Xin W, Wang C and Du F 2020 Boosting potassium-storage performance via the functional design of a heterostructured Bi2S3@RGO composite *Nanoscale* **12** 24394–402

[67] Shen C, Song G, Zhu X, Wang D, Huang L, Sun Z and Wu Y 2020 An in-depth study of heteroatom boosted anode for potassium-ion batteries *Nano Energy* **78** 105294

[68] Qin Y, Zhang Y, Wang J, Zhang J, Zhai Y, Wang H and Li D 2020 Heterogeneous structured Bi2S3/MoS2@NC nanoclusters: exploring the superior rate performance in sodium/potassium ion batteries *ACS Appl. Mater. Interfaces* **12** 42902–10

[69] Li K, Liu X, Qin Y, Zhao Z, Xu Y, Yi Y, Guan H, Fu Y, Liu P and Li D 2021 Sb2S3-Bi2S3 microrods with the combined action of carbon encapsulation and rGO confinement for improving high cycle stability in sodium/potassium storage *Chem. Eng. J.* **414** 128787

[70] Yang T *et al* 2020 Bi2Se3@C rod-like architecture with outstanding electrochemical properties in lithium/potassium-ion batteries *ACS Appl. Energy Mater.* **3** 11073–81

[71] Chen K-T, Chong S, Yuan L, Yang Y-C and Tuan H-Y 2021 Conversion-alloying dual mechanism anode: nitrogen-doped carbon-coated Bi2Se3 wrapped with graphene for superior potassium-ion storage *Energy Storage Mater.* **39** 239–49

[72] Zhang R, Bao J, Wang Y and Sun C-F 2018 Concentrated electrolytes stabilize bismuth–potassium batteries *Chem. Sci.* **9** 6193–98

www.ingramcontent.com/pod-product-compliance
Ingram Content Group UK Ltd.
Pitfield, Milton Keynes, MK11 3LW, UK
UKHW051852100926
8556IPUK00005B/581

9 780750 348690